SOTHIS PRESS

Sothis Press aims to completely reimagine the public domain book with an emphasis on universal access to captivatingly designed print editions. We believe that technology will enable every single book ever published to be perpetually accessible in beautiful print form. Sothis is beauty in ubiquity.

www.sothis.press
1 Waterhouse Square, London, ECIN 2ST

The Influence of the Blue Ray of the Sunlight and of the Blue Color of the Sky

In Developing Animal and Vegetable Life; In Arresting Disease and in Restoring Health in Acute and Chronic Disorders to Human and Domestic Animals

A. J. Pleasonton

SOTHIS PRESS

ISBN 978-1-0368-1007-8

The Influence of the Blue Ray of the Sunlight and of the Blue Color of the Sky is in the public domain.

This edition is published by Sothis Press The company is registered in England and Wales. The Company Number is 14900389.

THE

INFLUENCE

OF THE

BLUE RAY OF THE SUNLIGHT

AND OF THE

BLUE COLOUR OF THE SKY,

IN DEVELOPING ANIMAL AND VEGETABLE LIFE;
IN ARRESTING DISEASE, AND IN RESTORING HEALTH IN ACUTE AND CHRONIC DISORDERS TO HUMAN AND DOMESTIC ANIMALS,

AS ILLUSTRATED BY THE EXPERIMENTS OF

GEN. A. J. PLEASONTON, AND OTHERS,

Between the years 1861 and 1876.

Addressed to the Philadelphia Society for Promoting Agriculture.

"Error may be tolerated, when reason is left free to combat it."—Thomas Jefferson.
"If this theory be true, it upsets all other theories."—Richmond Whig.

PHILADELPHIA:
CLAXTON, REMSEN & HAFFELFINGER, PUBLISHERS.
1877.

PREFACE.

HAVING been much interested in the phenomena of the physics of the earth, the author, in offering to his readers a second edition of his work, "On the Influence of the Blue Color of the Sky in Developing Animal and Vegetable Life," may be indulged in his introduction into this preface of some views that his observations have led him to entertain relative to the variations of temperature, and changes of our seasons, which are in harmony with the subjects treated by him in this work.

The first edition of the following memoir was printed for distribution among scientific and literary institutions, and among persons of culture, for the purpose of attracting the attention of those for whom it was intended, to the subjects of which it treats. It was hoped that its publication would invite investigation into the nature, composition, and influences of those great forces which, in the poverty of our language, we call imponderables, that is to say, not to be weighed in the balance, and consequently never to be found wanting. This expectation is likely to be realized, if we may judge from the general interest that appears to be taken in the memoir, which has been manifested in the numerous applications that have been made to the author, from various parts of our country, for copies of it. The edition has now been distributed, yet so many persons who have applied for copies of the memoir are still without it, that it has been deemed advisable to issue another edition.

If, by a course of study, and observation of the great forces of nature, as they are exhibited, not in the laboratory, upon the minutest scale, but in those grand operations by which physical changes are at every moment developed before our eyes, we can succeed in penetrating the mysteries of their origin, of their evolution, of their application, and of their reciprocal conversions into each other, we shall become indeed wise in our generation, and mankind in the future will be able to rejoice in a development never yet reached in any preceding age.

By way of illustration of this idea, we may suggest that this planet is surrounded, at variable altitudes above its surface, by a canopy of cold, increasing in intensity with its distance above the earth. Now, we may ask, what produces the changes of our seasons? We answer, simply the descent or ascent of columns of this canopy of cold!

It has been observed, for any years, that the first frost of the autumn appears in Texas or Louisiana, or some other of the Gulf States, while at the same time no frost is observable in other localities situated much farther to the north—the commonly supposed place of departure of our winters. This frost, therefore, must come from the descent of the cold of the higher atmosphere immediately over the locality where it prevails. Following the valley of the Mississippi and those of its tributaries, frost appears successively in various places along those routes, till it reaches the vallies of the Northern Lakes, running along which it is felt in Northern New York and the New England States, and subsequently in the Middle and Southern Atlantic States. It does not reach the vicinity of Philadelphia until some fifteen or twenty days after it has shown itself on the Gulf of Mexico. Now would it not seem that the influences producing this frost are telluric, and not exclusively solar, as hitherto they have been supposed to be*

We know that in the ocean there are columns of fresh water which differ in temperature from the surrounding sea water, and with which they do not mingle for a long time. So is it for a hundred or more miles at sea, distant from the mouths of the great rivers Amazon, Orinoco, Mississippi, etc., whose fresh waters do not mix with the salt waters surrounding them, owing to the difference of their densities. In like manner the cold air of the upper atmosphere descends in columns of various extent over particular localities, to vary the temperature and change the seasons, on the surface of our earth, without mixing with the warmer and more expanded air beneath, which it displaces.

The spring and summer seasons are produced by increased radiations from the interior heat of the earth, forcing upwards the dense cold of winter, whose particles are so close together as to prevent the intrusion among them of the expanded warm air in its ascent. Much of the heat of the lower atmosphere is also developed in the conversion of vapor into clouds by condensation from cold.

It is in this way that our seasons are changed. Let our savans discover how and why these effects are produced. Until they do, it may be suggested that they are owing to electrical atmospherical disturbances in the upper atmosphere, repelling the negative electricity of those regions, and forcing the cold air to the surface of the earth, where it displaces the warmer and more rarefied and expanded air, and condenses in rain, snow and hail, the vapors it contains, driving the displaced warmer air to the tropics, and the heat from the tropics attracted to the condensed vapor in the clouds in the temperate zones to liquefy them in rain, producing winter.

In the opposite manner the warm seasons of spring and summer are produced by the positive electricity of the surface-air of the earth becoming warmed by increased radiation of heat from the interior of the earth, repelling itself, and displacing the upper strata of cold air, till by induction of electricity the temperature of the season is established.

Geologists tell us that in the early existence of this planet, the greater part of the earth's surface was covered with ice, and that this period of time is called the Glacial Period.

Let us imagine that the igneous action of the elementary substances of the interior of the earth's crust, just before that period, might have been so intense as by the radiation of its heat to the surface of the earth to rarefy the lower atmosphere, converting into vapor the water it contained, and forcing it upward till the whole surface of the earth was almost incandescent.

To restore the equilibrium, the canopy of cold repelled by its own negative electricity from above, which has been increased by the currents of polar electricity, largely developed by this central and interior igneous action—and attracted by the positive electricity in the heated atmosphere below—descended to the surface of the earth, condensing the vapors of the atmosphere into rain, and afterwards into hail and snow, driving the remainder of the warmer air of what we call, now, the temperate zones, to the tropics, and covering the surfaces of the earth, from the poles to the tropics, with a dense mantle of ice, freezing the rivers, bays, and seas of those latitudes. The internal central fires thus concentrated, in due season increased their radiation of heat, and melted the superjacent ice, which, breaking from the sides of glaciers in large masses, slid and rolled to the ocean, there becoming icebergs, and carrying with them those immense boulders which, torn from the mountain sides by the adhesion of the ice, have left the traces of their furrows on the slopes of the mountains, and have marked their courses till, by the melting of the bergs, they have been dropped in the ocean, which subsequently, by its subsidence, have left them dry on the land. If such was the cause of the glacial period, it would require no great stretch of fancy to comprehend the deluge of Deucalion or that of our great ancestor Noah, when the rain descended for forty days;

occasioned no doubt by a lesser descent of the canopy of cold (limiting its effect to the condensation of the vapors of the atmosphere into rain) than that which produced the glacial period.

If such effects follow from such causes, we need not be at a loss to account for the changes of our seasons, or the daily variations of temperature in every locality.

This edition of our memoir has been printed upon tinted paper with blue ink, as an experiment, in an attempt to relieve the eyes of the reader from the great glare, occasioned by the reflection of gas light at night from the white paper usually employed in the printing of books. If it shall succeed we may hope to see the tinted paper introduced for all books and periodicals.

PHILADELPHIA, July 29, 1871.

PREFACE TO THE LAST EDITION.

In the previous editions of my memoir "On the Influence of the Blue Colour of the Sky in Developing Animal and Vegetable Life," an erroneous impression has been created by the ambiguity of the language employed in describing the results of my experiments with light. From the tints reflected from the outside of the coloured glass, upon certain localities in my terraced garden, I fancied that the glass itself was of a violet tint, and so attributed the remarkable results within the grapery to violet rays. Upon my attention having been called to this apparent discrepancy, I investigated the matter, and found that the glass was of a dark mazarine blue—owing its colour to a preparation of cobalt, which had been fused with the materials composing the glass during its manufacture—and that the reflection of the violet ray on the outside was due to the irregular surface of the glass itself upon which the light of the firmament, as well as of the sunlight had fallen, and had been thus reflected. Whatever effect may be produced by the use of violet coloured glass is to be attributed to the proportion of the blue ray which enters into the composition of the violet rays of light, and not to those composite rays themselves.

This edition, begun in the summer of the year 1873, has been prepared at intervals snatched from the occupations of a busy life, which will account for any incoherences that may appear in the subjects as they are treated herein.

The following memoir was read by GEN. A. J. PLEASONTON. before the Philadelphia Society for Promoting Agriculture, on Wednesday, the 3d of May, 1871, at their room, S. W. corner of 9th and Walnut Streets, in the City of Philadelphia, upon the following request:

1309 WALNUT ST., *April 27th, 1871.*

MR DEAR GENERAL:

Will it suit you, and will you do us the favor to explain your process of using glass in improving stock to the Philadelphia Society for Promoting Agriculture, on Wednesday next, the 3d of May, at eleven o'clock, A. M., at their Room, S. W. corner of Ninth and Walnut Streets, (entrance on Ninth street)? You were kind enough to express to me. in conversation, your willingness to give us the result of your experiments.

Yours, very truly,

W. H. DRAYTON,

President.

GENERAL PLEASONTON.

Mr. President and Gentlemen of The Philadelphia Society for Promoting Agriculture.

At the request of my old friend and your respected President, I have attended your meeting this morning to impart to you the results of certain experiments that I have made within the last ten years in attempts to utilize the blue color of the sky in the development of vegetable and animal life.

I may premise that for a long time I have thought that the blue color of the sky, so permanent and so all-pervading, and yet so varying in intensity of color, according to season and latitude, must have some abiding relation and intimate connection with the living organisms on this planet.

Deeply impressed with this idea, in the autumn of the year 1860, I commenced the erection of a cold grapery on my farm in the western part of this city. I remembered that while a student of chemistry I was taught that in the analysis of the ray of the sun by the prism, in the year 1666, by Sir Isaac Newton, he had resolved it into the seven primary rays, viz: red, orange, yellow, green, blue, indigo and violet, and had discovered that these elementary rays had different indices of refraction; that for the *red* ray at one side of the solar spectrum being the least, while that of the *violet* at the opposite side thereof was the greatest, from which he deduced his celebrated doctrine of *the different refrangibility of the rays of light;* and further, that Sir John Herschel in his subsequent investigation of the properties of light had shown that the chemical power of the solar ray is greatest in the *blue rays*, which give the least light of any of the *luminous* prismatic radiations, but the largest quantity of solar heat, and that later experiments established the fact of the stimulating influence of the blue rays upon vegetation. Having concluded to make a practical application of the properties of the blue and violet rays of light just referred to in stimulating vegetable life, I began to inquire in every accessible direction if this stimulating quality of the blue or violet ray had ever received any practical useful application. My inquiries developed the facts that various experiments had been made in England and on the European continent with glass colored with each of the several primary rays, but that they were so unsatisfactory in their results that nothing useful came of them so far as any improvement in the process of developing vegetation was concerned. Finding no beaten track, I was left to grope my way as best I could under the guidance of the violet ray alone. My grapery was finished in March, 1861. Its dimensions were, 84 feet long, 26 feet wide, 16 feet high at the ridge, with a double-pitched roof. It was built at the foot of a terraced garden, in the direction of N. E. by E. to S. W. by W. On three sides of it there was a border 12 feet wide, and on the fourth or N. E. by E. side the border was only five feet wide, being a walk of the garden. The borders inside and outside were excavated 3 feet 6 inches deep, and were filled up with the usual nutritive matter, carefully prepared for growing vines. I do not think they differed essentially from thousands of other borders which have been made in many parts of the world. The first question to be solved on the completion of the frame of the grapery, was the proportion of blue or violet glass to be used on the roof. Should too much be used, it would reduce the temperature too much, and cause a failure of the experiment; if too little, it would not afford a fair test. At a venture I adopted every eighth row of glass on the roof to be violet colored, alternating the rows on opposite sides of the roof, so that the sun in its daily course should cast a beam of violet light on every leaf in the grapery. Cuttings of vines of some twenty varieties of grapes, each one year old, of the thickness of a pipe-stem, and cut close to the pots containing them, were planted in the borders inside and outside of the grapery, in the early part of April, 1861. Soon after being planted the growth of the vines began. Those on the outside were trained through earthen pipes in the walls to the inside, and as they grew they were tied up to the wires like those which had been planted within. Very soon the vines began to attract great notice of all who saw them from the rapid growth they were making. Every day disclosed some new extension, and the gardener was kept busy in tying up the new wood which the day before he had not observed. In a few weeks after the vines had been planted, the walls and inside of the roof were closely covered with the most luxurious and healthy development of foliage and wood.

In the early part of September, 1861, Mr. Robert Buist, Sr., a noted seedsman and distinguished horticulturist from whom I had procured the vines, having heard of their wonderful growth, visited the grapery. On entering it he seemed to be lost in amazement at what he saw; after examining it very carefully, turning to me, he said, "General! I have been cultivating plants and vines of various kinds for the last forty years; I have seen some of the best vineries and conservatories in England and Scotland, but I have never seen anything like this growth." He then measured some of the vines and found them forty-five feet in length, and an inch in diameter at the distance of one foot above the ground; and these dimensions were the growth of only five months! He then remarked, "I visited last week a new grapery near Darby, the vines in which I furnished at the same time I did yours; they were of the same varieties, of like age and size, when they were planted as yours; they were planted at the same time with yours. When I saw them last week, they were puny spindling plants not more than five feet long, and scarcely increased in diameter since they were planted—and yet they have had the best possible care and attendance!"

The vines continued healthy and to grow, making an abundance of young wood during the remainder of the season of 1861.

In March of 1862 they were started to grow, having been pruned and cleaned in January of that year. The growth in this second season was, if anything, more remarkable than it had been in the previous year. Besides the formation of new wood and the display of the most luxuiriant foliage, there was a wonderful number of bunches of grapes, which soon assumed the most remarkable proportions—the bunches being of extraordinary magnitude, and the grapes of unusual size and development.

In September of 1862 the same gentleman Mr. Robert Buist, Sr., who had visited the grapery the year before came again—this time accompa nied by his foreman. The grapes were then beginning to color and to ripen rapidly. On entering the grapery, astonished at the wonderful display of foliage and fruit which it presented, he stood for a while in silent amazement; he then slowly walked around the grapery several times, critically examining its wonders; when taking from his pocket paper and pencil, he noted on the paper each bunch of grapes, and estimated its weight, after which aggregating the whole, he came to me and said, "General! do you know that you have 1200 pounds of grapes in this grapery?" On my saying that I had no idea of the quantity it contained, he continued, "you have indeed that weight of fruit, but I would not dare to publish it, for no one would believe me." We may well conceive of his astonishment at this product when we are reminded that in grape-growing countries where grapes have been grown for centuries, that a period of time of from five to six years will elapse before a single bunch of grapes can be produced from a young vine—while before him in the second year of the growth of vines which he himself had furnished only seventeen months before, he saw this remarkable yield of the finest and choicest varieties of grapes. He might well say that an account of it would be incredible.

During the next season (1863) the vines again fruited and matured a crop of grapes estimated by comparison with the yield of the previous year to weigh about two tons; the vines were perfectly healthy and free from the usual maladies which affect the grape. By this time the grapery and its products had become partially known among cultivators, who said that such excessive crops would exhaust the vines, and that the following year there would be no fruit, as it was well known that all plants required rest after yielding large crops; notwithstanding, new wood was formed this year for the next year's crop, which turned out to be quite as large as it had been in the season of 1863, and so on year by year the vines have continued to bear large crops of fine fruit without intermission for the last nine years. They are now healthy and strong, and as yet show no signs of decrepitude or exhaustion.

The success of the grapery induced me to make an experiment with animal life. In the autumn of 1869 I built a piggery and introduced into the roof and three sides of it violet-colored and white glass in equal proportions—half of each kind. Separating a recent litter of Chester county pigs into two parties, I placed three sows and one barrow pig in the ordinary pen, and three other sows and one other barrow pig in the pen under the violet glass. The pigs were all about two months old. The weight of the pigs was as follows, viz: Under the violet glass, No. 1 sow, 42 lbs., No. 2, a barrow pig, 45½ lbs., No. 3, a sow, 38 lbs., No. 4, a sow 42, lbs., their aggregate weight 167½ lbs. The weight of the others in the common pen was as follows, viz: No. 1., a sow, 50 lbs., No. 2, a sow, 48 lbs., No. 3, a barrow big, 59 lbs., No. 4, a sow, 46 lbs; their aggregate weight was 203 lbs. It will be observed that each of the pigs under the violet glass was lighter in weight than the lightest in weight pig of those under the sunlight alone in the common pen. The two sets of pigs were treated exactly alike; fed with the same kinds of food at equal intervals of time, and with equal quantities by measure at each meal, and were attended by the same man. They were put in the pens on the 3d day of November, 1869, and kept there until the 4th day of March, 1870, when they were weighed again. By some misconception of my orders, the separate weight of each pig was not had. The aggregate weight of the three sows under the violet light on the 3d of November, 1869, was 122 lbs; on the 4th of March, 1870, it was 520 lbs., increase 398 lbs.

The aggregate weight of the three sows in the old pens on the 3d of November, 1869, was 144 lbs., and on the 4th of March, 1870, it was 530 lbs., increase 386 lbs., or 12 lbs. less than those under the violet glass had gained.

The weight of the barrow pig in the common pen on the 3d of November, 1869, was 59 lbs., and on the 4th of March, 1870, it was 210 lbs., increase 151 lbs. The weight of the barrow pig under the violet light, on the 3d of November, 1869, was 45½ lbs., and on the 4th of March, 1870, it was 170 lbs., increase 124½ lbs. The large increase of the weight of the barrow pig in the common pen is to be attributed to his superior size and weight on being put in the same common pen with the three sows, and which enabled him to seize upon and appropriate to himself more than his share of the common food.

If the barrow pig under the violet light had increased at the rate of increase of the barrow pig in the common pen, his weight on the 4th March, 1870, would have been only $161\frac{84}{100}$ lbs. instead of his actual weight of 170 lbs.—showing his rate of increase of weight to have been $8\frac{36}{100}$ lbs. more than that of the other barrow pig.

If the barrow pig under the sunshine in the common pen had increased at the rate of increase of the barrow pig under the violet glass, his weight on the 4th of March, 1870, should have been $224\frac{12}{100}$ lbs. instead of 210 lbs., his actual weight at that date.

By these comparisons it seems obvious that the influence of the violet-colored glass was very marked, although it must be borne in mind that owing to the great declination of the sun during the period of the experiment and the consequent comparative feebleness of the force of the actinic or chemical rays of the blue sky at that time, the effect was not so great as it would have been at a later period of the season; but the time of the experiment was selected for that very reason. The animals were not fed to produce fat or increase of size, but simply to ascertain, if practicable, whether by the ordinary mode of feeding usual on farms in this country, the development of stock could be hastened by exposing them in pens to the combined influence of sunlight and the transmitted rays of the blue sky.

My next experiment was with an Alderney bull calf born on the 26th of January, 1870; at its birth it was so puny and feeble that the man who attends upon my stock, a very experienced hand, told me that it could not live. I directed him to put it in one of the pens under the violet glass. It was done. In 24 hours a very sensible change had occurred in the animal. It had arisen on its feet, walked about the pen, took its food freely by the finger, and manifested great vivacity. In a few days its feeble condition had entirely disappeared. It began to grow, and its development was marvelous. On the 31st March, 1870, 2 months and 5 days after its birth, its rapid growth was so apparent, that as its hind quarter was then growing, I told my son to measure its height, and to note down in writing the height of the hind quarter, and the time of measurement—which he did. On the 20th of the following May (1870), just fifty days afterwards, my son again measured the hind quarter, and found that in that time it had gained *exactly six inches in height, carrying its lateral development with it.* Believing the question solved, the calf was turned into the barn-yard, and when mingling with the cows he manifested every symptom of full masculine vigor, though at the time he was only four months old. Since the 1st of April of this year, when he was just 14 months old, he has been kept with my herd of cows, and has fulfilled every expectation that I had formed of him. He is now one of the best developed animals that can be found any where.

These, gentlemen, are the experiments about which your curiosity has been excited. If by the combination of sunlight and blue light from the sky, you can mature quadrupeds in twelve months with no greater supply of food than would be used for an immature animal in the same period, you can scarcely conceive of the immeasurable value of this discovery to an agricultural people. You would no longer have to wait five years for the maturity of a colt; and all your animals could be produced in the greatest abundance and variety. A prominent member of the bar a short time since told me that his sister, who is a widow of a late distinguished general in the army, had applied blue light to the rearing of poultry, with the most remarkable success, after having heard of my experiments. In regard to the human family, its influence would be wide spread—you could not only in the temperate regions produce the early maturity of the tropics, but you could invigorate the constitutions of invalids, and develop in the young, a generation, physically and intellectually, which might become a marvel to mankind. Architects would be required to so arrange the introduction of these mixed rays of light into our houses, that the occupants might derive the greatest benefit from their influence. Mankind will then not only be able to live fast, but they can live well and also live long.

Let us attempt an explanation of this phenomenon. It is well known that differences of temperature evolve electricity, as do also evaporation, pressure suddenly produced or suddenly removed, in which may be comprised a blow or stroke, as, for instance, from the horseshoe in the rapid motion of a horse on a stone in the pavement, striking fire, which is kindled by the electricity evolved in the impact, or, again, from the collision of two silicious stones in which there is no iron, is electricity produced.

Friction even of two pieces of dried wood excites combustion by the evolution of hydrogen gas which bursts into flame when brought into contact with the opposite electricity evolved by the heat. Chrystallization, the freezing of water, the melting of ice or snow—every act of combination in respiration, every movement and contraction of organic tissues, and, indeed, every change in the form of matter evolve electricity, which in turn contributes to form new modifications of the matter which has yielded it.

The diamond, about whose origin so much mystery has always existed, it is likely, is the product of the decomposition of carbonic acid gas in the higher atmosphere by electricity, liberating the oxygen gas, converting it into ozone, fusing the carbon, and by the intense cold there prevailing, which is of opposite electricity, chrystallizing the fused carbon, which is precipitated by its gravity to the earth.

To the repellent affinity of electricity are we indebted for the expansive force of steam whose power wields the mighty trip hammer, propels the ship through the ocean, and draws the train over the land—and to the opposite electricities of the heated steam and the cold water introduced into the boiler to replenish it, do we owe those terrible explosions in steam boilers whose prevention has hitherto defied human skill. But the most interesting application of electricity, is in nature's development of vegetation. Let us illustrate it:

Seed perfectly dried, but still retaining the vital principle, like the seed of wheat preserved for thousands of years in mummy cases in the catacombs of Egypt, if planted in a soil of the richest alluvial deposits, also thoroughly dried, will not germinate. Why? Let us examine. The alluvial soil is composed of the *debris* of hills and mountains containing an extensive variety of metallic and metalloid compounds mingled with the remains of vegetable and animal matter in a state of great comminution, washed by the rains and carried by freshets into the depressions of the surface of the earth. These various elements of the soil have different electrical attributes. In a perfectly dry state no electrical action will occur among them, but let the rain, bringing with it ammonia and carbonic acid, in however minute quantities, from the upper atmosphere, fall upon this alluvial soil, so as to moisten its mass within the influence of light, heat, and air, and plant your seed within it, and what will you observe? Rapid germination of the seed. Why? The slightly acidulated, or it may be alkaline water of the rain has formed the medium to excite galvanic currents of electricity in the heterogeneous matter of the alluvial soil—the vitality of the seed is developed and vegetable life is the result. Hence vegetable life owes its existence to electricity. Herein consists the secret of successful agriculture. If you can maintain the currents of electricity at the roots of plants by supplying the acidulated or alkaline moisture to excite them during droughts, you will secure the most abundant and unvarying crops. To do this, your soil should be composed of the most varied elements, mineral, earthy, alkaline, vegetable, and animal matter in a state of greatly comminuted decomposition.

The poverty of soils arises from the homogeneous character of their composition. A soil altogether clayey, or composed of silicious sand, or the *debris* of limestone, or of alkaline substances exclusively, must necessarily be barren for the want of electrical excitement, which no one of the said elements will produce; but commingle them all with the addition of decomposed vegetable and animal matter, and you will form a soil which will amply reward the toil of the husbandman.

What do you suppose has produced the giant trees of California? Electricity! Since the west coast of America has been known to Europeans, and perhaps for centuries before, it has been subjected to the most devastating earthquakes. From the Straits of Magellan to the Arctic Ocean, traces of volcanic action are everywhere visible. Its mountains have been upheaved, broken, torn asunder, and sometimes, like Ossa upon Pelion, one has been superimposed on another.

All volcanic countries are noted in the temperate regions, where they produce anything, for the exuberance of their vegetable productions. Etna has been famous for its large Chestnut trees, which have given a name Catania to the town near its base.

The mineral richness of California has doubtless, by the *debris* of its mountains, carried into the valleys where grow these large trees, furnished an immense deposite of various matter which, under the favorable circumstances of the localities, have maintained for ages a healthful electrical excitement resulting through centuries of undisturbed growth in these vegetable wonders.

Who is there that has not been struck with admiration in looking upon the firmament, when the atmosphere was clearest, and was unclouded by the slightest vapor,—when, in the brightness of sunlight, it would put on its livery of blue, and display its resplendent and glorious beauties? How many myriads of mankind, in all ages, have gazed upon this magnificent arch, of what men call "sky;" and how few have ever asked the question, Why is the sky blue? and why should its intensity of blue vary in different latitudes, and in different seasons?

Humboldt said he had never seen its blue so intense as in the tropics and under the equator. Arctic navigators have also declared, that in the arctic regions the intensity of the blue color of the sky was amazing. Here are two extremes of latitude displaying the same effect; and in our own temperate region many have observed a variation in the intensity of the blue of the sky, in different seasons, extending from the early spring until the close of autumn, but never equaling in depth of color what is represented of it, either in the tropics or in the arctic or antarctic regions.

On no part of our planet is the development of vegetable life so grand, so various, so excessive and so constant as in the tropics and in the equatorial regions. While this wonderful display of vegetation is observed in these regions, the exuberance of animal life and the rapid growth of vegetable life in the arctic regions are said to be unequaled in any other part of our world. Let us see if these results in the two natural kingdoms may not be attributed largely to the same cause.

Recent discoveries have shown that the Zodiacal light over the equator and the auroræ borealis and australis are evolutions of electricity. In the arctic regions there is little doubt that the auroræ are constantly evolved, though they are not always visible. They have been seen to emerge from the surface of the ocean, at short distances from the observers, and ascending into the upper regions of the atmosphere, to present those corruscations of brilliant light, shooting as it were to the equatorial regions, in rapid flashes, for which they have been noted wherever observed.

In the equatorial regions it is well known that at certain periods of the year the accumulation of electricity in the upper atmosphere is so excessive, that the earth is shaken with thunderbolts, and the air illuminated by day as well as night with constant sheets of electric flame, as they rush with frightful velocity to their great centre of attraction, the earth and ocean in those regions. Whence does this electricity come, and where does it go?

If we may be permitted to form a conjecture, we might suggest that the sixty odd primary elements which enter into the composition of the crust of our planet—such as carbon, sulphur, phosphorus, oxygen, nitrogen, hydrogen, the metals, the metalloids, etc.—having been endowed by the Creator with different electrical qualities and conditions—when they were assembled together in this planet, evolved in the interior thereof electricity, light, heat, and magnetism in certain or variable qualities and quantities. These constitute the forces which in all probability cause the rotation of the earth upon its axis, and assist in its revolution around the sun. The electricity of the interior of the earth is supposed to be positive electricity —which, as soon as evolved there, would be repelled according to the law of electricity of the same character repelling itself —towards the poles of the earth, and escaping there, would be attracted by the negative electricity which surrounds the upper atmosphere, and would display itself by night as auroræ, corruscating toward the equator, to be there attracted by the heated equatorial regions, and descending to the earth, to be again absorbed by it, for further use. This escape of polar electricity into the upper atmosphere, and forming at night the auroræ, when visible, and by day the blue firmament or sky, will account for the intensity of the blue color of the sky both in the arctic regions and the equatorial regions.

This positive electricity of the central interior of the earth, repelling itself towards the poles, and from there into the atmosphere through the arctic and antarctic oceans, and attracted there by the negative electricity of the upper atmosphere, forms, by the union of the two electricities, the auroras, causing those crackling detonations heard during the prevalence of the most brilliant auroras, in high latitudes and evolving light, which, seen through the vaporous atmosphere of those latitudes, is displayed by refractions of its rays in the luminous corruscations of varying tints as the rays of the sun or moon are converted into the tints of the rainbow.

The negative electricity of those frigid regions attracted to the equator through the upper atmosphere is there concentrated in enormous quantities, which are conducted and discharged into the earth or ocean in the tropics, by the incessant fall of water in rain during the rainy seasons, every drop of water being a conductor of electricity, and every leaf of vegetation assisting in the conduct and distribution of this wonderful force into the earth.

As under certain circumstances electricity becomes magnetism, and this again is converted into electricity, we can comprehend how the auroral rays in some instances, following the law of dia-magnetism, are attracted in the northern hemisphere towards the southwest—magnetic currents flowing from east to west in opposition to the earth's motion from west to east; hence in the auroras you have rays shooting to the zenith over the equator, and others moving southwest, and others again due west.

The simultaneous appearance of auroras frequently observed in opposite hemispheres in corresponding latitudes would go to show their origin from a common impulse in the central interior repelling them towards the poles from under the equator.

We now come to a presumed explanation of one of the reasons for the blue color of the sky.

The sun's ray, or what is called the white light of the sun, was resolved by means of a glass prism, by Sir Isaac Newton, into the seven primary rays of light, viz., red, blue, violet, etc., and their combination again produced the white light--showing both synthetically and analytically of what the sun's light was composed.

It was announced in England about the beginning of this century, that the red ray of light was heating, the yellow ray was illuminating, *and the blue ray in a remarkable degree stimulated the development of vegetable life.*

From this discovery we can imagine the immense influence which the intensely blue color of the sky in the equatorial regions has and always has had in conjunction with the sun's white light, and the heat and moisture of those regions, upon the development there of vegetable life.

This intensely blue color of the sky in the arctic regions may also serve to explain the exuberance of animal life there. It being known that the deeper water of the arctic ocean is much warmer than the surface water which is often frozen, furnishes abundant food for its inhabitants. The increased temperature of this deep water is probably derived from radiation of heat from the interior of the earth under it—as all those regions are more or less volcanic; witness Iceland, Jean Mayer, Spitsbergen, etc. The laws of animal and vegetable life being very analogous, what would stimulate one would probably have a similar effect upon the other.

In the arctic waters you have warmth, food, light and electricity, escaping through the waters into the air, and all stimulating life.

Whoever has noticed the color of the electric spark in atmospheric air, from an electrical machine, will readily recognize its likeness of color to the blue color of the sky.

If experiments should be instituted to ascertain the electrical conditiou of the sky, *as associated with its blue color*, and they should satisfactorily establish the connection, the result would prove to be one of the greatest blessings ever conferred upon mankind. What strength of vitality could be infused into the feeble young, the mature invalid, and the decrepit octogenarian! How rapidly might the various races of our domestic animals be multiplied, and how much might their individual proportions be enlarged!

One of the most beautiful illustrations of the mighty influence of the blue color of the sky upon vegetation, is to be found in the green color of the leaves of plants. It is known that blue and yellow when mixed produce green, which is darker when the blue is in excess over the yellow, and the reverse when the yellow predominates. Now let us observe the process of germination. Seeds are planted in the soil—at first a white worm-like thread at the lower part of the seed appears; it is white, and contains all the primary rays of light; it is the root of the plant, and remaining under the soil continues white. At the upper end of the seed also appears a white swelling, which continues to grow upward till it approaches the surface of the soil, when a change occurs in its color. This is the leaf; it absorbs yellow from the soil which is brown (composed of yellow and black), and as it comes within the influence of the blue sky, it absorbs from it the blue light, which mixing with the yellow already absorbed, produces at first a yellowish-green, which finally assumes the deeper tinge of green that is natural to the plant. The plant blossoms, forms its seeds and seed-vessels, and having fulfilled its mission, the blue color of the leaves is eliminated, the leaves become yellow, and absorbing the carbon of the plant, they change their color to brown; the sap-vessels of the leaves are choked by the carbon; the leaves are dead and fall to the ground. Thus the blue ray is the symbol of vitality—the yellow ray that of decay and death.

Robert Hunt, in his Researches on Light, says "that the rays of greatest refrangibility, viz., the violet, &c., favor disoxygenation, but the rays of least refrangibility, viz., red, orange, &c., favor oxygenation."

"The experiments of Seunebier show that the most refrangible of the solar rays, viz., the violet, are the most active in determining the decomposition of carbonic acid gas by plants."

These experiments have been confirmed by Mr. Robert Hunt, who says, "that experiments have been made with absorbent media, and the light which has been carefully analyzed, permeating under the influence of *blue light*, in every instance oxygen gas has been collected, but not any under the energetic action of yellow or red light. * * It is only the green parts of plants which absorb carbonic acid: the flowers absorb oxygen gas. Plants grow in soils composed of divers materials, and they derive from these by the soluble powers of water, which is taken up by the roots, and by mechanical forces carried over every part, carbonic acid, carbonates and organic matters containing carbon. Evaporation is continually going on, and this water escapes freely from the leaves during the night when the functions of the vegetable, like those of the animal world, are at rest, and carries with it carbonic acid. Water and carbonic acid are sucked up by capillary attraction, and both evaporate from the exterir part of fhe leaves."

"There is no reversion of the processes which are necessary to support the life of a plant. The same functions are operating in the same way by day and by night, but differing greatly in degree. During the hours of sunshine the whole of the carbonic acid absorbed by the leaves or taken up with water by the roots is decomposed, all the functions of the plant are excited, the processes of inhalation and exhalation are quickened, and the plant pours out to the atmosphere streams of pure oxygen at the same time as it removes a large quantity of deleterious carbonic acid from it. In the shade the exciting power being lessened, these operations are slower, and in the dark they are very nearly, but certainly not quite, suspended."

"Although a blue glass or fluid may appear to absorb all the rays except the most refrangible ones, which have usually been considered as the least calorific of the solar rays; *yet it is certain that some principle* has permeated the glass or fluid which has a very decided and thermic influence. Numerous experiments have been tried with the seeds of mignonette, many varieties of the flowering pea, the common parsley, and cresses under the various tints of glass—with all of them the seeds have germinated, but except *under the blue glass* these plants have all been marked by the extraordinary length to which the stems of the cotyledons have grown, and by *the entire absence of the plumula*—no true leaves forming, the cotyledons soon perish and the plant dies; *under the blue glass* alone has the process gone on healthfully to the end."

"The changes which take place in the seed during the process of germination have been investigated by Saussure: oxygen gas is consumed and carbonic acid is evolved; and the volume of the latter is exactly equal to the volume of the former. The grain weighs less after germination than it did before; the loss of weight varying from one-third to one-fifth. This loss of course depends on the combination of its carbon with the oxygen absorbed, which is evolved as carbonic acid."

"For the discovery that oxygen gas is exhaled from the leaves of plants during the daytime, we are indebted to Dr. Priestley; and Seunebier first pointed out that carbonic acid is required for the disengagement of the oxygen in this process. M. Theodore de Saussure and De Candolle fully established this fact."

The experiments of Seunebier show that the most refrangible of the solar rays, viz., the violet, are the most active in determining the decomposition of carbonic acid by plants.

"We have now certain knowledge. We know that all the carbon which forms the masses of the magnificent trees of the forests and of the herbs of the fields has been supplied from the atmosphere, to which it has been given by the functions of animal life and the necessities of animal existence. Man and the whole of the animal kingdom require and take from the atmosphere its oxygen for their support. It is this which maintains the spark of life, and the product of this combustion is carbonic acid, which is thrown off as waste material, and which deteriorates the air. The vegetable kingdom, however, drinks this noxious vapor; it appropriates one of the elements of this gas—carbon—and the other—oxygen—is liberated again to perform its services to the animal world."

"The animal kingdom is constantly producing carbonic acid, water in the state of vapor, nitrogen, and in combination with hydrogen, ammonia. The vegetable kingdom continually consumes ammonia, nitrogen, water, and carbonic acid. The one is constantly pouring into the air what the other is as constantly drawing from it, and thus is the equilibrium of the elements maintained."

"Beccaria examined the solar phosphori, and ascertained that the violet ray was the most energetic, and the red ray the least so, in exciting phosphorescence in certain bodies."

"M. Biot and the elder Becquerel have proved that the slightest electrical disturbance is sufficient to produce these phosphorescent effects. May we not therefore regard the action of the most refrangible rays, viz., the violet, as analogous to that of the electric disturbance? May not electricity itself be but a development of this mysterious solar emanation?"

It has been long known to chemists that a mixture of chlorine and hydrogen gases might be preserved in darkness without combining for some time, but that exposure to diffused day light gradually occasioned their combination, and which is effected with the greatest speed by the *extreme blue and indigo rays.* M. Edmond Becquerel in 1839 first called attention to the "electricity developed during the chemical action excited by solar agency."

The experiments of Dr. Morichini, repeated by MM. Carpa and Ridolfi, that violet rays magnetized a small needle, were successfully confirmed by Mrs. Somerville.

"Light is not solely a radiant visible element. It has other properties which cannot be overlooked. It oxidizes, colors, bleaches. Light becomes absorbed—light changes into heat, and heat into electricity; in fact, light in its radiant visible character only shows one of its many phases. Light holds many forces within its beams. It has properties, powers of its own, which neither mathematician nor optician can grasp. It is a great chemical agent. Colors are produced by a change resulting from a polaric act of arrestation—yellow and red yellow belong to the acids; blue and red blue to the alkalies. The undulatory theory explains the radiant visible property of light, but it does not explain its chemical effects, the optical polarity of a chrystal and its connection with the polaric condition of its constituents—the diffraction, inflection, interferences, the oxidation of surfaces as the cause of natural colors, the presence of the chemical action of light, the presence of heat, electricity, magnetism; yet light produces all these phenomena; it vitalizes, and the organic action of light is witnessed in the fauna and flora around."

We have seen that blue light, and the violet ray which is a compound of it, and the red ray—being the most refrangible rays of the solar spectrum—excite magnetism,—and electricity, by which carbonic acid gas evaporated from growing plants is decomposed and oxygen thereof liberated to be absorbed again in maturing the flowers, fruit and seed of the plant, thus stimulating the active energies of the plant into its fullest and most complete development. Now this is just what I think is done in the vegetable world by the blue light of the firmanent. That blue light of the firmanent, if not itself electro-magnetism, evolves those forces which compose it in our atmosphere, and applying them at the season, viz., the early spring, when the sky is bluest, stimulates, after the torpor of winter, the active energies of the vegetable kingdom, by the decomposition of its carbonic acid gas—supplying carbon for the plants and oxygen to mature it, and to complete its mission.

In the experiment which I have made in the cultivation of grapes under violet light, I have endeavored to combine with it the blue light of the firmanent, causing the other rays of the solar spectrum to be absorbed while the blue and violet rays were permitted to permeate the violet glass into the grapery. The difference of temperature under the white glass and under the violet glass of the grapery is supposed to have excited currents of electricity sufficient to decompose more rapidly the carbonic acid gas that had been evaporated from the leaves of the vines, than would have been done under the influence of the sunshine alone—thus stimulating the increased absorption of oxygen, and the deposit of carbon in the vines, and constantly and quickly renewing the evaporation of carbonic acid gas. The result has been seen in the wonderfully large product of fruit, accompanied by a prodigious formation of new wood, to yield the crop of fruit for the ensuing year.

The investigations that have been made during the present century regarding light have developed the existence of some remarkable attributes; one of the most astonishing is the discovery that there is no heat *per se* in the sun's ray, though it is one of the causes which produce heat. This is established beyond dispute by the existence of the intense cold which prevails in the upper atmosphere, increasing with its altitude, and through which all the sunlight which reaches the earth must pass, but whose temperature it cannot alter. Hence you have at the present time the line of perpetual snow, according to Professor Agassiz, at an elevation of 15,000 feet at the equator, of 6,000 feet at the latitude of 45°, and gradually approaching the surface of the earth till it reaches it at 60° of north latitude, beyond which ice prevails nearly to the pole.

Aeronauts have remarked also at great altitudes above the earth that the thermometer had ceased to mark any variation of temperature when exposed in the full sunshine or in shadow.

A curious illustration of the fact that something more is needed than sunlight to produce heat is to be found in the fact stated by the famous arctic navigator, Dr. Scoresby, as well as by others, that when, after a long night in the arctic regions, the sun had appeared, though the thermometer was below 32° of Fahrenheit, and everything around was frozen hard, he observed that the pitch with which the seams of the planks of the ship had been payed, on the side of the ship exposed to the sun, was melted, notwithstanding the great declination of the sun and the small angle of incidence, that the nearly horizontal rays of it made as they fell upon the pitch, while that in the shade on the other side of the ship was so hard that it was with difficulty broken with a hatchet—other objects on the ship manifesting at the same time the low temperature marked by the thermometer. I am not aware that any explanation of this phenomenon has ever been attempted. I may, therefore, offer to suggest that the pitch being an electric or non-conductor of electricity and negatively electrified when the sun's ray positively electrified fell upon it, an explosion took place, heat was evolved, and the pitch was melted—thus proving that heat from sunshine is produced by the contact of an electricity opposed to that of the sun's rays.

As a corollary from what has just been stated, it may be observed that the heat of the equatorial and tropical oceans is not derived from the sun. We do not heat our houses by kindling fires at the tops of our chimneys or boil our water from above, but rather we descend into our cellars, and make our fires for that purpose in the furnaces constructed there. Besides, we know that from the surface of the water, if at rest, and from its many surfaces, if agitated by winds, the rays of the sun would be reflected in all possible angles corresponding to the angles of incidence of the rays themselves, and the heat would be lost in space. Whence comes, then, this ocean heat in the tropics, finding its vent in the arctic and antarctic regions through the Gulf Stream of the Atlantic, and the Japan Stream laving the shores of northeastern Asia, and the south-eastern current running along the south-western coast of South America to the Antartic seas? Does it not come by radiation from the interior of the earth from those great fires which, by the elastic gases and vapors engendered there, in many parts of the world upheave mountains and islands, and forming chimneys for themselves in their summits, belch out that superfluous heat, light, electricity, and magnetism which radiation to the surface of the earth at times is inadequate to discharge? And are not these great ocean currents of heated water merely channels or flues of radiation of heat from beneath, by which, for climatic purposes, the Omnipotent Creator has devised the means of distributing this interior heat over the surface of our planet?

All admit the existence of those great forces of nature in the interior of the earth, manifested through volcanic action in those imponderable elements of heat, light, electricity, and magnetism. Why are those forces there? May they not be the forces which turn the earth on its axis, and aid in propelling it around the sun? May not the frigid zones north and south furnish the cold cushions of water in the extreme depths of the ocean, of the uniform temperature of $39\frac{1}{2}°$ of Fahrenheit, and of nearly the greatest density known to that element, for the purpose of restraining and controlling the radiation of that great interior heat of the earth, which otherwise might be wasted?

Dr. Winslow, in his treatise on light, its influence on life and health, says: "Accurate calculations have been made as to the temperature of the ocean. The results obtained clearly establish that the lowest degrees of temperature are obtainable on the surface of the water; and that about ten feet below the surface the thermometer rises several degrees,—90° is said by Mr. Agassiz (son of Professor Agassiz,) to be the highest temperature he has known the ocean to attain; at very great depths of the ocean a uniform temperature of about 39½° has been found."

of the surface water of the ocean is attributable

warm ocean curre

d

Agricultural Bu-
June the sun

be inferr
latitude than at the equator; but that it should continue to increase
so clear. it is recol-
lected th quity of the
ould be less,

com

It strikes me that this explanation is not sound. I remember
day

doubt to a descending column of cold air from the upper at-

One of the observers in the northern district of Yukon states

midsummer, on the Upper Yukon, the only relief from the intense heat under which vegetation attains an almost tropical luxuriance, is the two or three hours during which the sun hovers near the northern horizon, and the weary voyager in his canoe blesses the transient coolness of the midnight air.' "

According to M. de Humboldt, the sky is bluer between the tropics than in the higher temperate latitudes, but paler at sea than in the interior of countries; the blue is less intense at the horizon than at the zenith. The early maturity of human life in the tropics is to be attributed to the stimulating influence of the enormous quantities of electricity, which, continually passing by day as well as by night in the auroras from the poles to the equator, and descending to the earth in those regions, in those dazzling sheets of lightning flame, so terrifying to all who have witnessed them, and conducted by the incessant rains prevailing there in certain seasons of the year—deoxygenate the enormous volumes of carbonic acid gas generated by the exuberaut vegetation, as well in its growth as in its decay, thus supplying excessive quantities of oxygen gas to stimulate and support the animal life, as well as carbon to the fresh vegetation which is being continually renewed—the circle of development and decay in the vegetable kingdom being thus always preserved.

We have thus seen that the magnetic, electric, and thermic powers of the Sun's ray reside in the violet ray, which is a compound of the blue and red rays. These constitute what are termed the chemical powers of the sunlight. That they are the most important powers of nature, there can be no doubt, as without them life cannot exist on this planet. Without these chemical powers there could be no vegetation. Without vegetation there could be no insect life, and no development of the higher order of animal existence. The earth would be without form and void, and we can now understand the potential meaning of the first sublime utterance of the Almighty in forming this earth, when he said "Let there be Light," and there was Light.

From the foregoing premises we deduce the following conclusions:

1. Heat is developed by opposite electricities in coujunction and in proportion to the quantity and intensity of those electricities in contact with each other, will be the intensity of the heat.

2. The blue color of the sky, for one of its functions, deoxygenates carbonic acid gas, supplying carbon to vegetation and sustaining both vegetable and animal life with its oxygen.

APPENDIX.

[I.]

UNITED STATES PATENT OFFICE. 119,242.

AUGUSTUS J. PLEASONTON, OF PHILADELPHIA, PENNSYLVANIA.

Improvement in Accelerating the Growth of Plants and Animals. Specifications forming part of Letters Patent No. 119,242, dated September 26, 1871.

To all whom it may concern:

Be it known that I, Augustus J. Pleasonton, of the city of Philadelphia, in the State of Pennsylvania. have discovered a new and valuable aid and improvement in accelerating the growth to maturity of plants, vines, vegetables, cereals, and the flora of the vegetable kingdom of nature, and of animals, fowls, fishes and birds of the animal kingdom of nature; and that I do hereby declare the following to be a full, clear, and exact description of the operation of the same by means of combining the natural light of the sun transmitted through transparent glass with the natural light of the sun transmitted through blue glass or any of the varieties of blue, as indigo or violet, in varied proportions of blue and white glass, from one of blue to eight of white, up to equal proportions of blue and white, as greater or less caloric is needed, according to the nature of the plants or animals, to accelerate their natural growth, increase their vitality, and hasten maturity; reference being also made to the accompanying drawing making a part of this specification, in which the figure represents one form of construction of a conservatory or grapery, in which A A A represent the clear or transparent glass, and B the blue or coloured glass. Proper ventilation is effected by means of wire cloth placed in the walls, as shown at C, and which can be opened and closed at pleasure by means of hinged glazed sashes, as shown at D. There is also represented at E a hinged sash, glazed with both clear and blue glass, for changing the angle of incidence to agree with the declination of the sun. These proportions of the natural light of the sun with the blue or electric transmitted rays may be varied to conform to the specific constitution of the varieties of life in the vegetable world and the varieties of constitution in the animal world, and can only be ascertained throughout both kingdoms by progressive and continued experiment. The proportions of the heating rays and the transmitted blue electric rays must be varied to conform to the constitutional vitality of the vegetable or animal, and care must be had that the heating or caloric light is not in excess of the electric or vitalizing and growing transmitted blue light.

I confine myself to no particular form, externally or internally, of the buildings to be used, whether they apply to the growth and propagation of plants, vegetables, fruits, &c., or to the growth propagation, &c., of animals, fishes and fowls; but the best form is that building which will receive the rays of the sun during its daily revolution as nearly perpendicular as practicable to the surfaces of the glass covering, so that the rays shall be as little deflected as possible, and the tiers or rows of blue glass, violet or other degrees of blue, shall be continuous over the entire portion of the building on which the sun shines, imparting in this way to every portion of the interior uniformly throughout the day the caloric and electric rays in the proportions of white and blue glass in their alternations. Such structures should be built on curves, conforming to the curve in which the sun moves in its daily revolution, and the alternating rows of white and blue glass should extend over the portions on which the sun shines, so that in the course of the day plants and vegetables, wherever they grow under the glass, will all have the same exposure to the caloric and electric transmitted light. Variations from these forms of buildings, and variations in the proportion of the natural caloric and blue electric light will, in degree, accelerate the growth and maturity of plants and animals depending upon their constitution and vitality; and the same proportions that hasten growth in the vegetable kingdom are not the best for many animals of the animal kingdom. Experience alone can determine the best proportions of natural and blue light, depending on the constitution of the animal and the nature of plants. In extreme northern latitudes the form given to the glass buildings so as to take the sun's rays perpendicularly to the surfaces during the day would vary from the form that should be given in southern latitudes to effect the same purpose. Therefore no one general plan for the construction of conservatories, graperies, houses for animals, &c., can be adopted or described beyond the rule for the builders to conform the shape of the glass portions so as to present their surfaces around his building in form to take the sun's rays as nearly perpendicularly as practicable, so as to avoid their deflection. All persons skilled in building will readily understand this principle, and be enabled to make use of the discovery and apply it to practical use, in whatever place he may live, extreme north or extreme south, within the limits of the sun's rising and setting. I prefer, as a transmitting medium for the electric rays of the sun, blue glass, violet and indigo; but I do not confine myself to the use of glass, as the sun's transmitted rays convey these colors through other media, producing in degree the same results.

In buildings for the treatment of invalids, whether they be men or animals, no particular form or construction of hospital, house or stable will be necessary, as the beds of invalid men and the places for animals can be so changed that the order of the means for transmitting the blue light may be very variable. The proportion of electric blue light and the natural light, however, should be constant, or as nearly so as practicable, after the proportions are ascertained by experience that prove most beneficial in their healing process.

I do not pretend to be the first discoverer of the vitalizing and life-growing qualities of the transmitted blue light of the solar rays, and its effect in quickening life and intensifying vitality.

I have found, upon patient and long experiments, running through many years, that plants, fruits of plants, vines and fruits of vines and vegetables so housed and inclosed as to admit the natural light of the sun through ordinary glass, and the transmitted light of the solar rays through the glasses of blue, violet or purple colours in the proportion of eight of natural light to one of the blue or electric light, grow much more rapidly, ripen much quicker, and produce much larger crops of fruit than the same plants housed and treated with the natural light of day, the soils and fertilizers and treatment and culture being identical in both cases and the exposure the same.

I have also found, by repeated and patient experiments of several years, that young animals, fishes and fowls under the same care, food, regimen, and treatment grow much more rapidly and to a much larger size under the influence of the combined natural light of day with the transmitted blue electric light than when exposed only to the natural sunlight, and that their flesh is equally good, and their health, vigor and constitutions are equal to those that, under the same circumstances of food, care and shelter, grow in the natural light. In these experiments with animals, fishes and fowls, I have not used the same proportions of natural light and transmitted blue light, viz: eight of natural to one of blue light, that I used in my experiments with vines, vegetables and fruits, but with the first named the proportions of natural and blue light were equal; and I prefer not those proportions of the natural caloric light and the transmitted electric light; yet I do not doubt that other proportions, depending upon the different organic constitutions in both the animal and vegetable creations, may be found to combine life-growing and vitalizing powers even exceeding the results I have produced, and still more productive of good in creating greater results. In these experiments I have discovered and proved that the transmitted blue light of the solar rays in its different degrees of intensity of color, in combination with natural sunlight, imparts vigour and vitality to the vegetation and life-growing principle in nature, heretofore unknown and never before utilized and applied to practical results of incalculable value to stock growing, to agriculture and horticulture, both as relates to time, labor and economy.

I have also discovered, by experiment and practice, special and specific efficacy in the use of this combination of the caloric rays of the sun and the electric blue light in stimulating the glands of the body, the nervous system generally, and the secretive organs of man and animals. It therefore becomes an important element in the treatment of diseases, especially such as have become chronic, or result from derangement of the secretive, perspiratory or glandular functions, as it vitalizes and gives renewed activity and force to the vital currents that keep the health unimpaired, or restores them when disordered or deranged.

Having thus fully described my discovery and invention, what I claim, and desire to have secured to me by Letter Patent, is

1. The method herein described for utilizing the natural light of the sun transmitted through clear glass, and the blue or electric solar rays transmitted through blue, purple or violet coloured glass, or its equivalent, in the propagation and growth of plants and animals, substantially as herein set forth.

2. The herein described construction of conservatories and other buildings, when the roof, walls or parts thereof are covered with alternating portions of clear and blue, purple, or violet glass or equivalents, as and for the purposes set forth.

In testimoney that I claim the above, I have hereunto subscribed my name in the presence of two witnesses at the city of Philadelphia, the 23d day of June, A. D. 1871.

AUGUSTUS J. PLEASONTON.

Witnesses:
H. Tunison,
H. A. Nagle.

[II.]

In the winter of the year 1872, I called at the Pennsylvania Hospital, on Pine street, between Eighth and Ninth streets, in this city, to suggest to its officers the introduction of my plan of using the associated light of the sun and the blue colour of the sky in alleviating the sufferings of, and probably in restoring to health many of their patients. On being presented to them, one of the resident physicians, on hearing my name mentioned, asked me if I was the author of the experiments with blue light of which he had read an account. On receiving my answer, he said; "I have something curious to tell you. I am a native of Brazil, where my father still resides; I have been educated in the United States; last week I received a package of books, pamphlets, &c., from my father, in Brazil, who had ordered them from Paris. In his accompanying letter my father directed my particular attention to a French pamphlet which detailed some remarkable experiments on animal and vegetable life, that had been made with blue glass and sunlight, that he thought would be useful to me in my medical profession. On examining the pamphlet I discovered it to be a translation in the French language of your memoir on that subject. The translator, however, had not mentioned your name in it, or even the name of the locality where the experiments had been made. It evidently was intended to convey the impression that the experiments had been made in Paris."

If the translator was a Frenchman we can pardon him for omitting the name of the author, in memory of the ancient Revolutionary alliance between his nation and our own. We can even condone his fault, smarting as he must have been under the then recent loss of Alsace and Lorraine—but we think that it might have occurred to him that the scene of my experiments was also the locality of the electrical experiments of Franklin, whom his countrymen and women always delighted to honor, and hence the name of Franklin's home might have been associated with the announcement of discoveries in physics that do no discredit even to those of Franklin himself.

[III.]

The Diamond; its Origin.—In former editions of this memoir I have attributed the origin of the diamond to electricity in the upper atmosphere decomposing carbonic acid gas, fusing the carbon, converting the oxygen gas into ozone, and crystalizing the fused carbon, under the great evaporating power of the intense cold there prevailing. The Atheneum says: "A somewhat novel idea is stated by M. Desdemaines Hugon, in a paper 'On the Diamond Diggings of South Africa,' which is printed in the *Revue Scientifique de la France et l'Etranger*. He states that the air is always highly electric where diamonds abound, and he intimates his opinion that this may throw some light on the formation of that gem."

[IV.]

[*From the President of the Indiana University.*]

INDIANA UNIVERSITY,
BLOOMINGTON, *June 15, 1871.*

GEN. PLEASONTON.

DEAR SIR:—I received a few days ago a pamphlet containing an account of your interesting experiments on the influence of the blue ray in developing animal and vegetable life. If the experiments, where it so difficult to determine the amount of influence due to the light, compared with that due to other circumstances, have been fairly made, as doubtless they have been, you have opened up a new field of great practical usefulness to all the world. Thanking you for your kindness in sending me your treatise, I remain,

Very respectfully yours,
T. A. WYLIE,

[V.]

[*From the President of the Lehigh University.*]

THE LEHIGH UNIVERSITY,
SOUTH BETHLEHEM, PA., PRESIDENT'S ROOMS, *July 10, 1871.*

MY DEAR GENERAL:

I have just received and at once read your very interesting paper *on violet rays, &c.*

The facts are astonishing, and your explanation evinces care, judgment and research.

I shall take pleasure in putting it among our scientific papers, and thank you for sending it.

Very faithfully yours,

HENRY COPPÉE.

GEN'L. PLEASONTON.

[VI.]

[*From the Hon. Wm. M. Meredith, late Secretary of the Treasury of the United States*]

MY DEAR PLEASONTON:

I have delayed thanking you for the pamphlet you sent me, till I should have read it, which I have now done twice, with very great interest and pleasure. I congratulate you sincerely on the discovery you have made, which must not only be greatly valuable in Agriculture and Horticulture, but in many other matters as well.

Always faithfully yours,
W. M. MEREDITH.

GEN. PLEASONTON, *Monday, 10th July, 1871.*

[VII.]

[*From Wm. A. Ingham, Esq., a Director of the Lehigh Valley Railroad Company.*]

320 WALNUT ST.,
PHILADELPHIA, *August 29th, 1871.*

DEAR GENERAL:

Allow me to return my thanks for the copies of your pamphlet. I have read it with great interest and am satisfied that your discovery will have wonderful results, revolutionizing in fact the science of horticulture.

I am, very truly yours,
WM. A. INGHAM.

GEN. A. J. PLEASONTON.

[VIII.]

[*From the Hon. Joseph R. Chandler, late Minister Plenipotentiary of the United States at the Court of Naples*]

153 NORTH TENTH STREET,
20th September, 1871.

DEAR SIR:

I thank you for a copy of the third edition of your pamphlet on "the influence of the blue colour of the sky." I cannot doubt the importance of your discovery, nor fail to see that the public must hold itself indebted to you for your interesting and successful experiments.

With great respect, your servant,
JOS. R. CHANDLER.

GEN. PLEASONTON.

[IX.]

DEPARTMENT OF THE INTERIOR,
PATENT OFFICE.

WASHINGTON, D. C., *August 15th, 1871.*

A. J. PLEASONTON, Philadelphia, Penn.

Your letter of the 14th inst., relative to your invitation to the examiner in charge of the Agricultural class of this office to call upon you to witness the influence of the "blue colour of the sky" in developing animal and vegetable life, is received.

In reply you are informed that Prof. Brainerd is at present confined to his room by sickness, but a leave will be given him for the purpose of accepting your invitation, as soon as he is able to travel.

Very respectfully, your obedient servant,
M. D. LEGGETT,
Commissioner.

[X.]

DEPARTMENT OF THE INTERIOR,
PATENT OFFICE.

WASHINGTON, D. C., *August 19th, 1871.*

DEAR SIR:

I have so far recovered from my late illness as to be able to pay you a visit in compliance with your invitation, for the purpose of examining your improvement in the construction of conservatories.

I purpose to leave this city on the 8 A. M. train on Tuesday, and shall therefore be due at Philadelphia at 1 P. M. * * * *

Respectfully,

J. BRAINERD,
Examiner.

GEN. A. J. PLEASONTON, Philadelphia, Pa.

[XI.]

DEPARTMENT OF THE INTERIOR,
PATENT OFFICE.

WASHINGTON, D. C., *September 6th, 1871.*

DEAR GENERAL:

Your drawing arrived this morning, and the patent will now go to issue, but will take the usual time.

The Commissioner yesterday introduced General Babcock, who is Superintendant of Public Grounds, and Consulting Engineer of the Board of Public Works. The object of his call was to learn particulars in regard to your *cerulean* process. I had a pleasant interview with him, at the close of which he desired me to write to you, asking the privilege of using your invention upon a grapery which he is now fitting up on the President's grounds. An answer directed to the care of myself or Commissioner of Patents, will reach him promptly. * * * * * *

Respectfully,

GENERAL A. J. PLEASONTON. J. BRAINERD.

[XII.]

PARIS, *September 29th, 1871.*

GENERAL PLEASONTON.

DEAR SIR:—I have just received and read with great pleasure, your very interesting paper from the *Gardener's Monthly*, of August last, concerning your experiments on the action of coloured light on plants and animals. You will find in the "Report of the

Department of Agriculture," at Washington, for 1869, a very long report of mine "on the influence of climalogic agents, atmospheric and terrestrial, upon agriculture," where, in the chapters of light and electricity, I have treated fully all these questions with a great number of experiments and quotations of authors. At that time I had no idea of any of your publications, although I had formed a bibliography on that subject of 1326 articles in every language. I am preparing a work in French and English on Agricultural Meteorology, and I should be most happy to mention in it your experiments, and to receive all that you have published. My name may be known to you through my papers on Meteorology at the French Academy and in America. I was the founder and director of the observatory at Havana until the beginning of our war, being now a victim of my patriotism. I correspond with several journals of the United States, as the *American Agriculturist*, the *Rural New Yorker*, etc., etc. * * * * * * * *

I remain, General, your most obedient servant,

ANDRÉ POËY.

54 Rue Mazarine, Hotel Mazarin.

[XIII.]

PARIS, *November 10th, 1871.*

GENERAL A. J. PLEASONTON.

DEAR SIR:—Your most affectionate of October 10th, is at hand, with seven copies of your interesting pamphlet. After a very careful study of that paper, I should advise you strongly to pursue your experiments on the influence of coloured lights on vegetable and animal life. There are still a great many points to be resolved, and, unfortunately, this important question has been totally abandoned in our days. Should you publish anything else, pray do not forget me. I shall be very happy to quote all your experiments in my works. At the next sitting of the French Academy, I shall also endeavor to have a little extract of your pamphlet inserted in the *Comptes Rendus* of that Institution, with a copy presented in your name, and also to M. Becquerel, M. Duchartre, the Meteorological Society, etc. I am waiting for the return of one of its perpetual Secretaries, M. Elie de Beaumont. I shall have the pleasure to send you whatever may be published on your experiments. I have sent another copy to the Meteorological Society of Vienna, very much interested in the study of periodical phenomena, treated in my second report to the Department of Agriculture. * * * * * * * * *

I remain your most obedient servant,

ANDRÉ POËY.

54 Rue Mazarin, Hotel Mazarin.

[XIV.]

PARIS, *November 24th, 1871.*

GENERAL A. J. PLEASONTON.

DEAR SIR:—As I had promised you I enclose the little extract presented to the French Academy of Science, Monday last, and which will appear to-morrow in the *Comptes Rendus.* I took particular pains to write a condensed letter, giving the most striking facts, to the perpetual Secretary, the great Geologist, M. Elie de Beaumont, who was very much interested in your experiments. A copy was also presented to the Academy, Becquerel Father, Duchartre, and Barral, the editor of the *Practical Journal of Agriculture,* who will reprint it in that paper. At the same time different scientific and political papers will make some mention of it. I shall send next week the translation of my letter to the excellent English journal called *Nature*; so your experiments and name will be, in short, spread through the scientific world in Europe.

* * * * * *

I remain, General, your most obedient servant,

ANDRÉ POËY.

54 Rue Mazarin, Hotel Mazarin.

[XV.]

[*From Rev. Henry A. Boardman, Pastor Tenth Presbyterian Church, Philadelphia.*]

MY DEAR GENERAL:

I thank you for your generous supply of the *Memoir,* and not less for the very kind terms of your note.

Once before in our national history the subject of "*Blue Light*" has caused a great commotion. There will be a *greater* still before long, and in a somewhat more beneficent direction. I heartily congratulate you on the just fame which is already assured to you as the reward of your great discovery.

I shall place the pamphlets where they will by appreciated.

Very sincerely yours,

H. A. BOARDMAN,

May 1st, 1872. 1311 Spruce Street.

[XVI.]

[*From the same.*]

1311 SPRUCE ST., *June 1st, 1872.*

MY DEAR GENERAL:

"*Laudari a viro laudato,*"—to be praised by a man who is himself praised,—the Latins used to think was a very nice thing. So

I take great pleasure in enclosing a letter from the *Rev. Dr. Sprague*, for forty years a pastor at Albany, one of the most accomplished and revered clergymen of our church or country, and enjoying a high European reputation. You will see what estimate he puts upon your great discovery, and how he prizes your *autograph*. For I took the liberty of sending him your kind note to me, for his famous autographic collection—the largest (some 200,000 specimens, I believe,) and finest in America.

I enclose, also, a note from Mr. Alex. Brown, Nineteenth and Walnut, to whom I gave the Memoir. I know it will gratify you. With sincere regard,

I am, dear General, yours,
H. A. BOARDMAN.

☞I design these two autographs for your collection, so you will not return them.

[XVII]

[*From Alexander Brown, Esq., Banker, &c.*]

PHILADELPHIA, *May 30th, 1872.*

REV. H. A. BOARDMAN.

DEAR SIR:—I thank you for the copy of Gen. Pleasonton's address before the "Philadelphia Agricultural Society."

I have read it with great interest, and think that the successful result of his experiments of the blue colour on animal and vegetable life must carry conviction to every mind.

Very respectfully, yours,
ALEX. BROWN.

[XVIII.]

[*From the Rev. Dr. W. B. Sprague, an eminent divine of Albany, New York.*]

FLUSHING, *May 30, 1872.*

MY DEAR DR. BOADMAN.

Since I wrote you yesterday, (I believe misdating my letter,) I have read the pamphlet you kindly sent me, with astonishment and admiration. I am not chemist enough to pronounce upon every part of it, but it seems to me that the man who could have written it is destined to be a great benefactor to the world; I do not see why it should not mark the introduction of a new and better era. I shall lay it away, with the author's autograph, as containing everything concerning him that I should desire.

With much love, as ever, yours,
W. B. SPRAGUE.

[XIX.]

[*From H. A. Boardman.*]

1311 SPRUCE ST., *Oct. 10th.*

MY DEAR GENERAL:

We are all prepared to testify that the *blue glass grapes* are in size, color and flavor of the very choicest. If there be gainsayers send them to us. We give you many thanks for so generous a sample of your crop. And what *bunches*, too!

The fresh testimonies you recite are very remarkable—a further presage of the certain and early attention which will soon be given to this whole subject, by men of science. I regret that I am compelled to send this bare acknowledgment of your extremely interesting letter.

I am, very truly yours,
H. A. BOARDMAN.

[XX.]

[*From Lieut. Col. Charles Manby, Royal Volunteer Engineer Staff Corps, England.*]

60 WESTBOURNE TERRACE, HYDE PARK.
LONDON, *March 23d, 1872.*

MY DEAR GENERAL:

Pray accept my thanks for your kind letter of 5th inst., and for the six copies of your most interesting paper, which I shall distribute to the persons most capable of comprehending it, and of repeating the experiments here.

I am grieved to say that my dear old friend A. H., of Copenhagen, for whom I ventured to ask you to send me your paper, has died in the interval, and he never received your paper, nor the copies of the *Comptes Rendus de l'Academie de Sciences de Paris*, treating of the subject, of which I procured him exemplaires. His son will, I hope, keep up his Horticultural Experiments and when I next go to Denmark I will tell you whether any experiments have been tried of your system. I have many friends who will, I think, try the system, and if you desire to make it known, send me some more copies and they shall be well placed among influential persons. I am a member of the Horticultural and the Botanical Societies of London, and in my capacity of Honorary Secretary of the Institution of Civil Engineers of England, I am in communication with scientific men, so I can make your system well known to everybody. It is most interesting as an investigation, and I will try and get it tried in every way. * * * * *

Believe me, my dear General, yours sincerely,
CHARLES MANBY.

[XXI.]

[*From J. T. Alden, of Newport, Kentucky.*]

NEWPORT, KENTUCKY, *May 26th, 1872.*

GENERAL A. J. PLEASONTON, PHILADELPHIA.

DEAR SIR:—Your esteemed favor of 23d inst., with pamphlets, at hand, for which please accept my sincere thanks.

I read your treatise with absorbing interest and satisfaction, and was amazed at the wonderful discoveries evolved by your critical observations and the scientific deductions and logical conclusions, and still more astounded by their grand and overwhelming demonstrations.

Your mind and vision have penetrated into the labyrinth of the "imponderable" deep of nature, and eliminated from her secret chambers great practical truths that hitherto have been buried in an abyss too profound for even man's comprehension. My dear sir, I do most sincerely congratulate you as the author of a discovery ranking in great practical value with those of Morse, Newton, Fulton and Watt. I cannot feel you will soon be adequately rewarded, because truths like these are too abstruse for immediate apprehension by the common mind. But time will reduce your grand theory (no longer theory in your hands,) to practice, and not until then will your great efforts be duly appreciated.

I bow in deep grateful devotion to you, as the chosen instrument of God in communicating this valuable knowledge to mankind; and may it be your happiness, as of Morse, to see its adoption by your fellow-creatures throughout the civilized world, as well as like him to reap the full fruition of its honors and fruits. And if you are not deluged with letters, I would like to ask if these principles may be applied in a more humble way than your demonstration exhibits? Can they be made subservient to the production of early vegetables by the ordinary hot bed cultivation? Have you experimented "on this line," or has any one under your instruction? If two or three weeks can be anticipated over the hot bed culture now so common, it would equal 400 miles of latitude in influence and results. On this point, if consistent with your time and convenience, I should be highly gratified to hear from you, stating any knowledge in your possession of results or discoveries in this line of industries.

With considerations of profound respect, I am, dear sir,

Your obedient servant,

J. T. ALDEN.

I will confer with you touching the area of territory desired hereafter.

[XXII.]

[*From Dr. John C. Brown, late Professor of Botany in the South African College, Cape Town, Cape of Good Hope, Africa.*]

PHILADELPHIA, 3405 BARING STREET,
16th October, 1873.

SIR:—I have tried to procure a copy of your valuable treatise on blue light as an organic stimulant, but have failed. May I ask you where I can procure a copy? May I ask if you have collected any information on the climatic effects which have followed the extensive destruction of forests in America, and if you can inform me where I may procure information on this point? I have just completed the tour of the forest districts of central and northern Europe, collecting information for the use of the Government at the Cape of Good Hope, and having come to the Conference of the Evangelical Alliance, I am desirous of such information, and I shall feel greatly obliged if you can supply it. I leave for Pittsburgh on Monday. Address me to the care of REV. O. EASTMAN, 311 West Twenty-Ninth Street, New York.

My permanent address is, REV. D. BROWN, Berwick-on-Tweed, England.

I am, sir, respectfully yours,

JOHN C. BROWN, LL. D., F. R. G. S., F. L. S., &c.

Late Professor of Botany in the South African College, Cape Town.

TO GENERAL PLEASONTON.

Mr. President and Gentlemen of the

Philadelphia Society for Promoting Agriculture:

It is now more than three years since I had the honour to read before you my memoir "on the influence of the blue colour of the sky in developing animal and vegetable life, as illustrated by certain experiments I had instituted and continued between the years 1861 and 1871."

The subject was so entirely novel, and the results of the experiments were so surprising, that men were lost in amazement when they contemplated the facts as they were narrated, and began to conjecture the bearing that these facts were destined to have upon the comfort, the health and the prosperity of mankind.

As a knowledge of the experiments and the conclusions deduced from them became diffused, various criticisms appeared in many journals, some of which were humorous, and intended to be facetious; others treated the subject with grave dignity, not knowing exactly what to make of it; while others, again, grasping it in its important relations, as by intuition, welcomed it as a long step in advance in the knowledge of the great truths in physics which mankind are so anxious to acquire. All this was perfectly natural. The little knowledge which men have has been acquired by great labour, industry, privation, and perhaps through a long course of arduous study. They are, therefore, loath to abandon preconceived notions upon any subject. It would be a loss of so much mental capital. A new idea, therefore, upon any familiar subject naturally excites doubt, and is met with disapproval until, by a free and full discussion, its merits are understood, when, if it is established by facts and conclusive reasoning upon them, it is accepted as sound, though it may displace all preëxisting notions in opposition to it.

Such has been the history of the publication of my memoir, and of the wonderful discovery that it describes. I proceed now to communicate to you some facts in connection with this subject, which are very curious, instructive and important.

It may be remembered that in the month of May, 1871, a great hailstorm visited this city and neighbourhood, and inflicted immense damage among gardens, green houses, &c. Among the sufferers was Mr. Robert Buist, Sr., in his extensive glass houses, near Darby, in some of which nearly all of the glass was broken. The damage was promptly repaired, and the houses reglazed as before, with colourless glass. After which, my memoir on the influence of the blue colour of the sky, &c., which had been read before your society in the beginning of May, of that year, was printed and published. It was then too late for Mr. Buist to introduce blue glass into his forcing houses—but fully informed of the results of my experiments he adopted an expedient, which differing somewhat from my experiments confirms the conclusions thereon to which I had arrived, and which will prove a valuable addition to our appliances in horticulture.

Mr. Buist had at this time a very large and valuable collection of geraniums which had become diseased; some of them had died, others were feeble, losing their leaves and flowers, and others again, though blooming, were sensibly being deprived of the brilliant tints of colour which characterized their several varieties.

It occurred to Mr. Buist that if he should paint with a *light blue colour* the inside surface of each pane of glass in one of his houses, leaving a margin of an inch and a quarter in width of the glass in its uncoloured condition all around the painted surface on each of the panes of glass, and then place his sickly geranium plants in the house under this glass so painted, the vigour of his plants might be restored.

The experiment was made, and was successful. The plants began to revive soon after they had been placed in this house. In two days thereafter they began to put forth new leaves, and at the end of ten days their vigour was not merely restored, but Mr. Buist assured me that the plants he had thus treated were more healthy and vigourous than he had ever seen similar plants of the same varieties to have been. Their colours were not only restored but their tints were intensified.

During the summer of 1871, Mr. Dreer, one of our most successful horticulturists, called my attention to another confirmation of my theory, which had just come to his notice. It was as follows, viz.:

A professional gardener in Massachusetts (near Boston) had been trying for several years to protect his young plants, as they were germinating, from various minute insects which fed upon them, sometimes as soon as they were formed. For this purpose he adopted nearly every expedient of which he had any knowledge, and even used the primary rays of sunlight separately. Nothing succeeded, however, in these experiments but the blue ray, which proved itself to be a perfect protection against the attacks of these insects. He made a small triangular frame, similar in form to a soldier's tent, covered it with blue gauze, such as ladies use for their veils. Having prepared a piece of ground, he sowed his seed in it, and, covering a portion of the ground thus prepared with his little blue frame and gauze, he left the other parts exposed to the attacks of the insects. His plants outside of this frame were all eaten by the insects, as soon as they germinated, while those under it escaped entirely from their depredations. This experiment was tried many times, and always with similar results.

This gardener had written an account of his experiments to Mr. Dreer, and had forwarded to him one of his small blue gauze frames, in order to its introduction here to the attention of our gardeners. This was shown to me by Mr. Dreer, with the gardener's account of his experiments with it.

The explanation of this phenomenon, I think, is this. The sunlight negatively electrified in passing through the meshes of the blue gauze of the frame, which is positively electrified, excites an electro-magnetic current sufficiently strong to destroy the feeble vitality of the eggs or of the insects themselves, which are in the soil with the seed, leaving the seed to germinate more rapidly under its influence. One remarkable circumstance in these experiments was that the combination of sunlight with blue light, while it destroyed these noxious insects injurious to vegetation, at the same time stimulated the development of the growth of the plants it had preserved.

Having introduced blue glass into the windows of the sleeping apartments of my servants in one of my country houses, it was observed that large numbers of flies, that had previously infested them, were dead soon after its introduction, on the inside sills of the windows. This effect seemed to be produced by a like cause to that on the insects injurious to vegetation as described by the gardener of Massachusetts in his experiments. Various experiments have been made in several parts of this country as well as in Europe, with this associated light, in developing vegetable life according to my suggestions and with results corresponding to those that I have obtained. A lady of my acquaintance, residing in this city, informed me that having some very choice and rare flowering plants in pots in her sitting room, which were drooping and manifesting signs of disease, she threw over them a blue gauze veil, such as ladies wear, and exposed them to the sunlight, when she was highly gratified to discover that in a very short time they were fully restored to health and vigour.

A gentleman in West Philadelphia having a large lemon tree, which he prized highly, placed it in his hall near to the vestibule door, the side lights of which were of glass of different colours, blue and violet predominating; the sunlight passing through these side lights fell upon a portion of the branches of this lemon tree; great vigour was imparted thereby to the vitality of these branches, which were filled with very fine lemons, while the other branches of the tree that did not receive the light from these blue and violet panes of glass, were small, feeble and apparently unhealthy, and were without fruit.

It will be remembered that during our late civil war, when commercial intercourse between the Northern and Southern States had ceased, the sale of early fruits and vegetables in the markets of the principal northern cities, was monopolized by their producers in the states of New Jersey and Delaware, and on the eastern shore of Maryland. This was a very valuable trade, and enriched many of those engaged in it. The price of land in these regions became enhanced in value, and the people resident there enjoyed unusual prosperity. On the restoration of peace all this was changed; the people along the Atlantic slope of Virginia, North and South Carolina and of a part of Georgia, at once entered upon the cultivation of fruits and vegetables for the northern cities, and owing to their lower latitudes and earlier seasons, and improved modes of cultivation, they have secured their lost markets, and are now rapidly recovering from the effects of the war. All this, of course, is a corresponding loss to the farmers of New Jersey, Delaware and the eastern shore of Maryland, and as a consequence the value of farming lands in these sections has been sensibly depreciated. A large portion of this trade can be recovered by the application of my discovery to the cultivation of vegetables and fruits, and their maturity can be hastened so as to equal that of those of the Southern States herein referred to.

The early vegetables used in my family are, for the most part, started in pots under blue and plain glass, then transplanted into proper soil, and are ready for use several weeks before I could otherwise obtain them. As an illustration, we have been using on my table since July 12th, of this year, Stowell's evergreen sugar corn, grown in this way, while I am informed that it is one of the latest in the season to mature; it will be at least two weeks later than now, August 10th, before any of it grown otherwise in the ordinary course of growth will be ready for use.*

As it is only the very early and very late vegetables and fruits that remunerate the grower, while the abundance of the regular crops reduces the prices oftentimes below cost, it is truly the interest of all persons engaged in furnishing such foods to mankind, to produce them and sell them when the prices are highest, viz., at the beginning and end of their seasons.

Cotton and tobacco, in the Middle States, can be raised and matured according to this process, so as to avoid entirely the September frosts, and to compete in yield and quality with any of the cottons grown in the Southern States, unless it may be the Sea Island cotton. I have myself raised and matured cotton plants on my lawn in this city, year after year, which produced as fine and large bolls as I have ever seen in Carolina or Georgia, and this without the use of blue glass, and before I had made my discovery of its wonderful influence on vegetation.

A machine has been invented and patented at Washington City, by which a man, with it and a mule, can set out in a day growing cotton plants which would cover an immense area of land. Now if these plants are started according to my directions, under these glasses, and then transplanted into suitable soil after the spring frosts are over, the heat and moisture of the summer in the Middle States, which probably are in excess of those of the Southern States at that season, will rapidly ensure the maturity of the plants; and crops can be thus raised which will compete favorably with those of any other

---

* The above was written in 1874.

section of the country. This same principle of hastening the maturity of plants, applies with still greater force to higher latitudes where the seasons of growth are necessarily short.

It is estimated that people residing six or eight degrees of latitude farther north than the present latitude of cultivation of various plants, may be enabled to enjoy many plants and fruits of which they are now deprived, by the introduction of the process of development that I have herein sketched.

What boundless blessings may not be obtained in this manner for the populations of Northern Germany, Southern Russia, of Scandinavia, Northern China and even the Steppes of Tartary, and some parts of Siberia which may be brought within the influence of this wonderful power, and thus, by increasing the comforts of life, hasten the progress of their civilization. So much for vegetation and what may be done with it. We will now invite your attention to the stimulating influence exerted by this associated blue and sunlight upon animal life.

An esteemed friend of mine, of high character, Commodore J. R. Goldsborough, of the United States Navy, having been assigned to the command of one of our western naval stations in the latter part of the year 1871, caused some experiments to be made with the associated blue light of the firmament, and sunlight, and subsequently addressed to me a letter, of which the following is a copy, viz:

MOUND CITY, ILLINOIS, *May 31st, 1872.*

To GENERAL A. J. PLEASONTON, *Philadelphia, Penn'a.*

GENERAL:—Presuming that it would be agreeable to you to learn the results of some experiments that I caused to be made, after having read the pamphlet you did me the honor to place in my hand, "On the Influence of the Blue Color of the Sky, in Developing Animal and Vegetable Life," I proceed to detail them to you: The first experiment was made here by the Surgeon of this station, who, having had every alternate pane of uncoloured glass removed from each of two windows in his parlour, and having substituted for them corresponding panes of blue glass, proceeded to place a number of plants and vines of many varieties, in pots, in the room so as to receive the associated light of the sun and the blue light of the firmament upon them.

In a very short time the plants and vines began to manifest the effects of the remarkable influences to which they had been subjected. Their growth was rapid and extraordinary, indicating unusual vigour, and increasing in the length of their branches from an inch and a half to three inches, according to their species, every twenty-four hours, as by measurement.

The second experiment was made in a comparison of the development of the newly hatched chickens of two broods of the same variety. In each of these two broods were thirteen chickens, all of which were hatched on the same day.

Comfortable but separate quarters near to each other were assigned to the two broods, with their respective mothers, on the lawn; one of the coops, containing a hen and her brood, was partly covered with blue and plain glass; the other coop, also containing a hen and her brood, did not differ from the coops commonly used in this country.

The chickens of each brood were fed at the same times and with equal quantities of similar food. Those under the blue glass soon began to display the effects of the stimulating influence of the associated blue and sunlight by their daily almost visible growth, increase of strength and activity, far exceeding in all these respects, the developments of the chickens of the other brood which were exposed to the ordinary atmospheric influences.

I will also relate to you what I imagine to be another remarkable circumstance having relation to this subject.

On the 29th of January, 1872, the wife of one of the gentlemen on the station gave birth prematurely to a very small child, which weighed at the time only three and a half pounds. It was very feeble, possessing apparently but little vitality. It so happened that the windows of the room, in which it was born and reared, were draped with blue curtains, through which and the plain glass of the windows, the sunlight entered the apartment. The lacteal system of the mother was greatly excited, and secreted an excessive quantity of milk, while at the same time the appetite of the child for food was greatly increased, to such an extent indeed, that its mother, notwithstanding the inordinate flow of her milk, at times found it difficult to satisfy its hunger.

The child grew rapidly in health, strength and size; and on the 29th of May, 1872, just four months after its birth, when I saw it, before I left Mound City, it weighed twenty-two pounds.

Whether this extraordinary result was the effect of the associated blue and sunlight, passing through the curtains and glass of the windows, or not, I do not profess to determine, but I give you the facts of the case, which are in complete harmony in their developments with the results of the experiments on domestic animals that you yourself have made. With great regard,

I remain, very truly, yours,

JOHN R. GOLDSBOROUGH.

It will be seen from this statement that this child had grown eighteen pounds and a half in four months, or four and five-eighth pounds per month, and considering its apparently slight hold upon life, at its birth, we may unite with the Commodore in believing it to be "a remarkable circumstance."

On the 15th February of this year, 1874, two newly born lambs, one weighing three and a half pounds, the other weighing four pounds, were taken from their mothers and placed in one of the pens on my farm fitted with blue and uncoloured glass; they had not received any nourishment from their dams, they were fed alike, and without any design to increase largely their weight, with skimmed cow's milk. When they were three months old, they were weighed—one of them weighed fifty-one pounds, the other fifty-five pounds—at two weeks old their teeth were so much developed that they began to eat hay.

The flesh of lambs is deemed to be a delicacy. From this experiment, it would appear that in three months from birth two lambs have gained forty-seven and a half and fifty-one pounds respectively, which, at the market price of forty cents per pound, would yield in one case twenty dollars and forty cents, and in the other twenty-two dollars, for the lambs weighing respectively fifty-one and fifty-five pounds.

Farmers who raise domestic animals for food have here a very simple and inexpensive process by which their gains may be very largely increased.

A gentleman of my acquaintance having a canary bird that had been a very fine singer, was surprised to discover that, without any apparent cause, the bird had ceased to sing, refused to eat, and evidently was in a declining state of health, and it was feared that he would soon die. I recommended the owner to try the effect of blue and sunlight upon the bird. He consented. The cage was removed with the bird to the bathroom of the owner's house, whose windows contained variegated glass, blue and violet in excess. The cage, with its occupant, was suspended so that the sunlight passing through these lights might fall upon the cage. The bird began to recover very soon, its appetite returned, and in a little while its song, which its owner assured me, was sweeter, stronger and more spirited then he had previously known it to be.

At the close of the late civil war in this country, I bought a pair of mules that had been used in the military service of the government. A little while after the purchase it was discovered that one of them was completely deaf, having had his hearing destroyed by the noise of heavy firing during the battles in which he had been employed. Thereupon I directed the teamster who had charge of him, to be particularly careful in using him, and to treat him with great gentleness and kindness on account of his infirmity. Two or three years after he came into my possession, this mule was seized with acute rheumatism of so violent a character that the poor animal could not walk. Before this time he, with other animals, had been removed to a new stable that I had built, in which he was kept for several months without being used for work. He gradually got better of his rheumatism, but his deafness continued until this spring, when he recovered entirely both from his deafness and rheumatism. Over each of the doors of this stable I had caused to be placed a transom, with panes of blue and colourless glass therein. The stall of this mule was before a door with such a transom over it. When the the sun arose in the morning, he cast his light through this transom on the neck and top of the head of this mule. Before he set in the afternoon he threw his light again upon the head and neck of this mule, through the transom of another door on the northwestern side of the stable; the effect of this light upon the animal has been the cure of his rheumatism, and the removal of his deafness. He is now as healthy and hearty a mule as you will see anywhere. The removal of this deafness was produced by an electro-magnetic current, evolved by the two lights upon his auditory nerves and exciting them to healthy action.

These last two incidents just mentioned, serve to introduce the subject of the influence of the associated blue and sunlight upon animal health and particularly upon Human Health.

It is known that silk is one of the most important staple products of Italy. It is also known that much of the high prices which this staple product bears in commerce, is due to the difficulty experienced in hatching and rearing the silk worms which produce the cocoons or balls on which they wind the silk drawn from their bodies. To hatch the eggs of the silk worm, an even temperature of a certain degree of heat is indispensable, and great care in feeding and keeping them clean is required after the worms are hatched.

An eminent Italian chemist, after the publication of the results of my experiments with blue light, instituted some experiments in the rearing of the silk worms. He placed a certain number of the eggs that produce the worms under plain glass, of which, in the hatching and rearing, 50 per cent. died. He then placed the same number of eggs under violet glass, of which only 10 per cent. perished. Had he used blue glass in his experiments it is probable that the loss would have been nearly nominal. As the rearing of silk worms for the European factories has become an important industry in California, we may expect great success will follow the efforts to raise them, when the stimulating influence of blue light shall be applied properly.

While we are considering this subject, it may be as well to allude to the vitalizing influence of the associated blue and sunlight of this discovery in the cure of human and other animal diseases, and I may mention here a most extraordinary case in which its power was manifested.

In the latter part of August, 1871, I chanced to visit a physician of this city, of my acquaintance, whom I found to be in great distress, and plunged in the lowest despondency. On inquiring the cause, he told me that he feared that he was about to lose his wife, who was suffering from a complication of disorders that were most painful and distressing, and which had baffled the skill of several of the most eminent physicians here, as also of others of equal distinction in New York. He then stated that his wife was suffering great pains in the lower part of her back, and in her head and neck, as also in her lower limbs; that she could not sleep; that she had no appetite for food and was rapidly wasting away in flesh; and that her secretions were all abnormal. I said to him, "Why don't you try blue light?" to which he replied, "I have thought of that, but you know how it is with wives; they will frequently reject the advice of a husband, while they would accept it if offered by any one else. This has deterred me from recommending blue light, but I think that if you should recommend it to her she will adopt it, for she has great confidence in your judgment." I told him that I would most certainly recommend it to her. Accordingly we went up to her sitting room in the second story of the main building, having a southern exposure, the house being on the southern side of the street. We found her seated at an open window, the thermometer up in the nineties; she was looking very miserable, greatly emaciated, sallow in complexion, indicating extreme ill health, and her voice very feeble. On inquiring of her relative to the state of her health, she described it very much as her husband, the doctor, had done. When I had put to her the same question I had proposed to her husband, viz: "Why don't you try blue light?" "Oh!" she replied, "I have tried so many things, and have had so many doctors that I am out of conceit of all remedies; none of them have done me any good; I don't believe that anything can relieve me." To which I remarked, "Nonsense! you have many years of life yet remaining, and if you will try blue light you will live to enjoy them." To which she answered, "Are you in earnest? Do you really think that blue light would do me any good?" "Certainly!" I said, "I do, or I would not recommend it to you; my experience with it fully justifies my opinion." She then said she would try it, and asked me how it should be applied. I then told her and her husband in what manner the application of blue light in her case should be made, and how often and when it should be repeated, and they both promised that the trial with it should be made the next day.

Six days after this interview I received a note from the doctor, asking me to send him some copies of my memoir on blue light, &c., which he wished to forward to some of his distant friends, and at the close of it he had written: "You will be surprised to learn that since my wife has been under the blue glass, her hair on the head has begun to grow, not merely longer, but in places on her head where there was none new hair is coming out thick." This was certainly an unexpected effect, but it displayed an evident action on the skin, and so far was encouraging. Two days after the receipt of this note I called to see the doctor, and while he was giving me an account of the experiment with the blue light, his wife entered the office, and coming to me, she said, "Oh, general! I am so much obliged to you for having recommended to me that blue light!" "Ah!" said I, "is it doing you any good?" "Yes," she said, "the greatest possible good. Do you know that when I put my naked foot under the blue light, all my pains in the limb cease?" I inquired, "Is that a fact?" She assured me that it was, and then added, "My maid tells me that my hair is growing not merely longer on my head, but in places there which were bald new hair is coming out thick." She also said that the pains in her back were less, and that there was a general improvement in the condition of her health.

Three weeks afterwards, on visiting them, the doctor told me that the arrangement of blue and sunlight had been a complete success with his wife; that her pains had left her; that she now slept well; her appetite had returned, and that she had already gained much flesh. His wife, a few moments afterwards, in person, confirmed this statement of her husband, and he added: "From my observation of the effects of this associated blue and sunlight upon my wife, I regard it as the greatest stimulant and most powerful tonic that I know of in medicine. It will be invaluable in typhoid cases, cases of debility, nervous depressions, and the like." It was at this time that the first symptoms in the improved condition of the health of the Prince of Wales, who had been dangerously ill in England, were announced, when the doctor added: "Now, in this case of the Prince of Wales, could he have been submitted to this treatment with the associated blue and sunlight baths, his recovery would be in one-tenth part of the time that it will take under the usual treatment."

I introduce here a copy of the letter that I received from this physician, Dr. S. W. Beckwith, on this subject. It is as follows, viz.:

"ELECTRICAL INSTITUTE, 1220 Walnut street,
"PHILADELPHIA, *September* 21, 1871.

"*To General A. J. Pleasonton.*

"MY DEAR SIR:—In following out the suggestions from you at our late conversation concerning the application of the associated blue light of the sky and sunlight for the cure of debility and nervous exhaustion, I have found some very singular results.

"The application of your theory to the cultivation of plants and the development of animal life, has been wonderfully successful; but it will, in certain conditions of human suffering, prove to be a far greater blessing to mankind, if judiciously used. As an illustration, I offer the following facts, viz:

"My wife had been suffering from nervous irritation and exhaustion, which resulted in severe neuralgic and rheumatic pains, depriving her of sleep and appetite for food, and producing in her great debility, accompanied by a wasting away of her body, and changing the normal character of her secretions.

"I had prepared a window sash fitted with blue glass, which was inserted in one half of one of the windows in her sitting-room. The sash of the other half of the same window was fitted with uncoloured glass, the window having a southern exposure, and receiving, from ten and a half o'clock A. M. till four o'clock P. M., the full blaze of the sun's light. The shutters of the other window (there being two windows in the room) were closed, excluding all light from it, and light was also excluded from the upper sash of the first mentioned window.

"This arrangement I found to furnish too strong a blue light for my wife's eyes; and, besides, it was not in accordance with your instructions. So I introduced an equal number of panes of clear glass and of blue glass into the sash, and then my wife exposed to the action of these associated lights those parts of her person which were the subjects of her neuralgia. In three minutes afterwards the pains were greatly subdued; and in ten minutes after having received the lights upon her person, they almost entirely ceased for the time being, whether they were in the head, limbs, feet, or spine. With each application of the sun and blue light bath, *relief* was given immediately. There is no doubt in my mind that in cases of exhaustion from long-continued fevers and other debilitating causes, the application of this principle that you have discovered will restore the patients to health with a rapidity tenfold greater than can be effected by any other treatment within my knowledge.

"Congratulating you upon your grand discovery, as well in science as in animal Hygiene,

"I remain, very truly yours,

"S. W. BECKWITH.

"P. S.—From a close examination of the effects of these associated lights of the sun and the firmament, I am of the opinion that they furnish the greatest stimulant and the most powerful tonic that I am acquainted with in medicine.

"Very truly yours,

"S. W. BECKWITH."

About this time (September, 1871), one of my sons, about 22 years of age, a remarkably vigourous and muscular young man, was afflicted with a severe attack of sciatica, or rheumatism of the sciatic nerve, in his left hip and thigh, from which he had been unable to obtain any relief, though the usual medical as well as galvanic remedies had been applied. He had become lame from it, and he suffered much pain in his attempts to walk.

I advised him to try the associated sun and blue light, both upon his naked spine and hip, which he did with such benefit that at the end of three weeks after taking the first of these baths of light, every symptom of the disorder disappeared, and he has had no return of it since—a period now of three years.

Some time since two of my friends, Major Generals S—— and D——, of the United States regular army, were on duty in this city. On making them a visit at their official residence, I saw on the window-ledge as I entered the room, a piece of blue glass of about the size of one of the panes of glass in the window. After some conversation, General D. said to me, "Did you notice that piece of blue glass on our window-ledge?" I said, "I had observed it." "Do you know what it is there for?" To which I replied, that "I did not!" He then said, "I will tell you—S. and I have been suffering very much from rheumatism in our fore-arms, from the elbow-joints to our fingers' ends; sometimes our fingers were so rigid that we could not hold a pen—we have tried almost every remedy that was ever heard of for relief, but without avail; at last I said to S., suppose we try Pleasonton's blue glass, to which he assented—when I sent for the glass and placed it on the window-ledge. When the sun began about ten o'clock in the morning to throw its light through the glass of the window, we took off our coats, rolled up our shirt sleeves to the shoulders, and then held our naked arms under the blue and sunlight; in three days thereafter, having taken each day one of these sun-baths for 30 minutes on our arms, the pains in them ceased, and we have not had any return of them since—we are cured."

It is now more than two years since the date of my visit to these officers. Two months ago General S. told me that he had not had any return of the rheumatism, nor did he think that General D. had had any—General S. in the meantime had been exposed to every vicissitude of climate, from the Atlantic Ocean to Washington Territory, on the Pacific, and from the 49th degree of north latitude to the Gulf of Mexico, and General D. was then stationed in the far North.

In the beginning of March, 1873, I was called upon by Mr. Henry H. Holloway, a very respectable gentleman, doing business in this city as a bookseller, who came to consult me on the subject of his mother's illness, and to ask my opinion in regard to the propriety of using blue and sunlight baths in her case. He stated that his mother had been confined to her bed for more than two months, and that she was suffering excruciating pains in her head, spine and other parts of her body; that she could not bear to be moved in bed; that she could not sleep, and having no appetite, she was rapidly wasting away in flesh and strength; that her physician had not been able to make any impression upon her malady, and that the family were in despair lest she should die; that its members had been summoned to her bedside that afternoon to see her probably for the last time, and if I thought that these blue and sunlight baths would relieve his mother, he wished to have them tried. From his account it was evident that her situation was critical, and that there was a serious disturbance of the electrical equilibrium in her system; I told him very frankly that I thought his mother could be greatly benefited by the use of the said baths of light, and I informed him how and how often these baths of light should be administered. He expressed himself much gratified by my explanations and said, that he would urge his mother and her physician to give them a fair trial. I received from him subsequently a letter, of which the following is a copy, viz:

"PHILADELPHIA, *April 14th, 1873.*

"*To General A. J. Pleasonton.*

"DEAR SIR:—Knowing that you have been assiduously investigating the curative properties of blue light (for human diseases) for several years past, a feeling of gratitude prompts me to take the liberty of communicating a few facts that may be of some interest to you.

"About six weeks since I heard you explaining to an acquaintance of yours, the way in which blue light should be arranged in windows, so as to take sun-baths thereby. In enumerating the classes of invalids that would be benefited by such baths, you mentioned those afflicted with spinous or nervous diseases.

"I was an interested auditor; for my mother, Margaret C. Holloway, residing in Chesterfield township, Burlington county, New Jersey, had then been confined to her bed for about two months, her entire nervous system being apparently incurably affected. It was probably a regular consumption of the nerves. She appeared to be wasting away very rapidly, and we had but little, if any, hope of her recovery.

"At my request, after first obtaining the full consent of herself and the attending physician, blue window lights (purchased from French, Richards & Co., of this city,) were suitably arranged in the west windows of her room, the east windows being too much shaded by trees to admit the light properly. During the first week thereafter, the weather was so unfavorable that only one sun-bath could be taken; but the next week, three or four were taken on consecutive days.

"From the commencement of her sickness, she had not been able to sit up more than a few minutes each day, just while the nurse made the bed; but in a few days after the several sun-baths were taken in succession, she surprised the entire family by getting up and dressing herself while they were at breakfast. She probably over-exerted herself as she was not so well for two or three days thereafter. However, she continued to improve very rapidly, and has now almost or entirely regained her usual health.

"I may just here state the most important perceptible effects of the sun-bath.

"During most of the time of her illness, mother suffered from an intense pain in the upper part of the spine and in her head, and the galvanic battery had been frequently and regularly used in the hope of mitigating it. The sun-baths relieved this pain very materially; and also induced a profuse perspiration that relieved the interior organs from their obstructions, and which relief medicines, as well as the galvanic battery, had failed to produce.

"These are the important facts in the case.

"The attending physician would probably maintain that the remedial virtue was mainly or altogether in his medicines, but the circumstances are such as to induce the belief that mother's speedy recovery was in a great degree attributable to the curative properties of the blue glass. I am so fully convinced of this that I shall hereafter use the glass in a similar way, in all cases of protracted sickness in my own family, whenever practicable.

"Very respectfully yours, &c.,

"HENRY H. HOLLOWAY,

"No. 5 South Tenth street, Philadelphia, Pa."

This lady soon afterwards recovered her usual good health, and on its re-establishment, she made several visits to her sons residing here. In two of these visits, I had the pleasure to see her. In one of the interviews that I had with her, she told me that for two years prior to the use of these baths of light she had had no perceptible perspiration, but that after the third of these light baths, a most copious perspiration broke out all over her person, but particularly profuse on her neck and shoulders, and that she had called her daughter to witness it, who scraped it with her hands from her neck and shoulders as a groom does from a horse that has been hard driven or ridden in summer. She dates her recovery from the restoration of her power to perspire, which she attributed to the effect of the associated sun and blue lights.

I addressed a note to the attending physician in this case, asking from him a statement of the case, with its diagnosis, &c. From his reply I make the following extract, viz: "Mrs. H. had been sick some two or three weeks with excessive spinal iritation amounting to partial paralysis of the right side, with intense neuralgia from the occiput down to the foot, including the right arm. This condition was greatly improved before the blue glass was used. She was almost free from pain, but nervous iritation remaining at this time I made use of the galvanic battery, which she thought done her a great deal of good.

"I think it was some two or three days after that, the blue light was used. She says that she took it about twelve times altogether, from a quarter to a half hour each time.

"You can draw your own conclusion, if there was any benefit derived from blue light.

"My dear sir, I would not have you imagine that I do not have any faith in your theory, for I confidently believe that it has a most powerful influence, both on the animal and vegetable kingdoms.

"I should like, at some future period, to give it a fair trial; consequently, if it would not be encroaching too much on your time, I should like very much to hear from you in regard to your experience of its application and result, the manner and mode by which it may be used, and should there be any benefit derived by its use, I would most cheerfully transmit that fact to you.

"Respectfully yours,

"J. G. L. WHITEHEAD.

"CROSSWICKS, *April 2d, 1873.*"

I have introduced here the extract from the letter of Dr. Whitehead merely to show the desperate condition of his patient, her agonizing suffering, and the well founded apprehensions of the patient's family—that the situation of the patient was extremely critical, and fully justified the use even of experiment with a new practice, in the attempt to relieve her. When they saw that the expedients resorted to during her long sickness had failed to produce the desired results, Dr. Whitehead, himself, is stated by Mr. Holloway to have given his full consent to have the experiment with the blue light made in the case of Mrs. Holloway, she also desiring it, which is conclusive that she had not been so much benefited by his treatment of her as to wish to continue it longer, and that he also was in doubt as to its efficacy from the adoption of another practice.

About this time, Mr. H. H. Holloway, the gentleman whose mother's case is given above, being a great sufferer from rheumatism, from which he had been unable to obtain relief, determined to try in his own person the efficacy of the sun and blue light bath, and after having tested it to his entire satisfaction, addressed me a letter, as follows, viz:

"PHILADELPHIA, *October 17th, 1873.*

"*Gen. A. J. Pleasonton.*

"DEAR SIR:—In the spring of 1872, I was afflicted with the rheumatism (sciatica,) for nearly two months, and I suffered from a recurrence of the same, at intervals, until last spring. At that time the surprising effect which your blue glass sun-baths produced in restoring my mother to health (an account of which I sent you a few months since,) induced me to try the same for the rheumatism.

"I took three or four such baths of sun and blue light, in accordance with your directions, and have had no returns of the rheumatism since, although six months have now elapsed; and I have been much exposed in stormy weather. My limbs have been a little stiff, but without pain, two or three times during long continued storms, which was probably owing to the mercury contained in the drugs taken by me, when first attacked in 1872.

"I have deferred writing to you on the subject for several months, so that sufficient time might elapse to be sure of the permanence of the effect of the blue glass sunbaths.

"I am fully confident that a fair trial of said sunbaths will seldom if ever fail to cure the rheumatism, and I wish that so simple and inexpensive a curative agent may speedily become popularized.

"Very respectfully,

"HENRY H. HOLLOWAY.

"No. 5 South 10th street, Phila."

In the further consideration of this subject, I introduce here some extracts from a letter received from Dr. Robert Rohland, a distinguished physician residing in New York.

"NEW YORK, *July 13th, 1873.*

"*General A. J. Pleasonton.*

"SIR:—Dr. McL. told me, three days since, that you had written to him about a new edition of your highly interesting pamphlet on blue light that you were preparing, that would contain additional results that you had obtained in your experiments with blue light as a healing power. I can readily believe in its efficacy, and I very much regret that I have been unable to continue my own experiments in the same direction, by which many new facts would have been developed in all likelihood to the great benefit of suffering humanity. Be that as it may, you deserve the warmest thanks for having extended your experiments so far, making the professional physicians to feel ashamed that none of them thought it worth their while to draw practical consequences from your experiments in the development of animal and vegetable life. As the effect of blue light is identical with '*od-force*' it might be of interest to you to hear of some surprising phenomena produced on sensitive persons in connection with blue light and corroborating the results of '*od-force*' and '*odified preparations.*'

"1. Compare with your results of the blue light on the Alderney bull calf the statement of Dr. Henry B. Heind, page 36 of my pamphlet on '*od-force*,' case No. 17, and you will find the similar surprising growth of babies, by using my '*od-magnetic* sugar of milk.'

"2. I exposed, about a year ago, a man suffering with severe rheumatism to the influence of the blue light through two glass panes. He felt, after fifteen minutes, much relieved, and could move about without pains, but complained of a nasty metallic taste on his tongue. The same happened to a friend who visited me during odo-magnetizing sugar of milk, when I placed his hand in the blue and violet rays of the prism.

"Dr. Hardis, assistant physician of Dr. E. B. Foote, has the same *metallic* (copper) taste, whenever he takes some of my odo-magnetic sugar of milk, on his tongue; also Dr. Fincke, a highly educated and reliable physician in Brooklyn, who experimented a great deal with od-force produced by the blue and violet rays of the prism, and who placed the hand of a man within these rays, and the latter complained of having a taste like verdigris on his tongue.

"These examples show that the blue and violet light and the od-force generated in this way are of an electric positive nature; and it is very much to be regretted that Professor Von Reichenbach reversed the poles, and, in his works, calls this pole, which is analogical in its effects to the *positive pole* of any electric or electro-magnetic apparatus, the 'odic-negative one,' causing by that uselessly an unavoidable confusion."

In the latter part of March, 1874, I received a letter from Major-General Charles W. Sanford, late the commander of the National Guard of the city of New York, of which the following is a copy:

"462 West Twenty-Second street,
"NEW YORK, *March 29th*, 1874.

"*To Major-General Pleasonton*,
"918 Spruce street, Phila., Pa.

"GENERAL:—Will you oblige me with a copy of your pamphlet upon the use of blue glass? I had some time since an opportunity to read it, and having an invalid daughter, her physician was induced to try the experiment of having blue glass inserted in her windows. She has been materially benefited by its use, and I am anxious to investigate the subject.

"She has also a number of plants in her sitting-room, which have grown and flourished in an extraordinary manner under its influence. I am, General, very respectfully,

"Your obedient servant,
"CHARLES W. SANFORD."

Extract from a letter of Dr. Robert Rohland, of New York, received by me in June, 1874.

"NEW YORK, *June* 28, 1874.

"*To General A. J. Pleasonton*,
"Philadelphia.

"SIR:— . . . . . . Several gentlemen have made some experiments with blue light under my direction, with very favourable results, especially Dr. L. Fisher, in a case of general debility and exhaustion, and Dr. McLaury, in a case of very troublesome tumor.

"Very respectfully yours, truly,
"DR. ROBERT ROHLAND."

Extract from a letter of Dr. Wm. M. McLaury, of New York, received by me in August, 1874.

"*To General Pleasonton*, Phila.

"DEAR SIR:—Understanding through Dr. R. Rohland that you are about to publish a new edition of your article on the blue ray, with some additional matter, I suppose that you would like to hear of my experience therewith.

"I regret to state that my experience is as yet very limited, but I have great hopes that by extensive experiments, with careful observation, we will yet find it to be an important agent in combating disease.

. . . . . . "In a little girl, one month old, was found a hard resisting tumour about the size of a robin's egg, in the sub-maxillary region of the left side. I had it placed in such a position that the rays of light through a blue glass should impinge upon it one hour, at least, each day. This tumefaction disappeared entirely within forty days.

"The child has developed astonishingly; is now seven months old; is exceedingly bright and happy; has not known an hour's sickness or discomfort. Its peculiar freedom from infantile ills I attribute, at least in some degree, to the influence of the blue light.

"With great respect, yours,

"WM. M. McLAURY.

"New York City, *August* 20*th*, 1874."

Some time since, Mrs. C., the wife of Major-General C., a distinguished officer of the United States regular army, told me that one of her grandchildren, a little boy about eighteen months old, had from his birth had so little use of his legs that he could neither crawl nor walk, and was apparently so enfeebled in those limbs that she began to fear that the child was permanently paralyzed in them.

To obviate such an affliction, she requested the mother of the child to send him, with his two young sisters, to play in the entry of the second story of her house, where she had fitted up a window with blue and plain glass in equal proportions. The children were accordingly brought there and were allowed to play for several hours in this large entry or hall under the mixed sun and blue light. In a very few days, Mrs. C—— told me that the child manifested great improvement in the strength of its limbs, having learned to climb by a chair, to crawl and to walk, and that he was then as promising a child as any one is likely to see.

In the case of the child, whose premature birth occured at the naval station at Mound City, in Illinois, Commodore Goldsborough was informed by its mother, a short time since, that it had continued to improve in health, size and vigour, since the Commodore had last seen it, and that it was then a perfect specimen of infantile development.

The case of this child, described by Commodore Goldsborough, is a very remarkable one, for, having been prematurely born, it may be presumed that its organization was not as completely developed as it would have been had it fulfilled the entire period of its gestation—and consequently it would seem that the association of the blue and sun light had repaired all the deficiencies in its organisms existing at its birth.

We have, in these instances that I have advanced, manifestations of the remarkable variety of powers as developed in the several cases, all differing from each other in their various disorders, and all having been restored to their normal condition of health and vigour; and, in some instances, having had that condition increased and intensified.

We have had moribund flowering plants, not only arrested in their course of decay, but reinvigourated, and their beautiful tints of colour greatly improved.

We have had branches of a tropical fruit tree, that were exposed to the action of blue light, made highly fruitful, while others of the same tree, not similarly exposed, bore no fruit, and were feeble and apparently unhealthy.

We have an immature infant child, defective in its developments at its birth, made perfect in its parts, and strengthened so as to become a striking instance of infantile health, vigour and beauty.

We have had in another infant child, only one month old, an obstinate tumour to be absorbed, and a degree of bodily vigour imparted to it that defied the attacks of all infantile disorders after the tumour had disappeared.

We have had poultry of the same variety, hatched on the same day, presenting such different stages of advanced development, after the lapse of the same period of time, to those of similar poultry reared in the common way, that incredulity must yield to well established fact, and surprise give way to conviction.

We have had the vocal powers of a singing bird, that had ceased to sing, again excited, and its musical tones again poured forth with greater force, richness and beauty than it had before ever displayed, to the delight of all who have heard it.

The deaf has been made to hear: in a domestic animal, the mule, which for nearly ten years, and perhaps longer, had heard not at all; and the stiffness of his limbs with rheumatism has given way to the natural elasticity of his normal condition of health. Under this most potent influence, lambs that may be used for the food and clothing of man, have been so greatly developed in so short a time that we may reasonably hope that the rearing of domestic animals for food may be so largely extended and improved, that immense numbers of mankind who, from the costliness of such food heretofore, had never tasted it, may, in the near future, be no longer deprived of the use of this most stimulating and nourishing article of flesh diet.

But the greatest value of this application of blue light, will be found to be in its curative power in human and animal disorders of health.

In the cases before quoted in the human family, rheumatism, both chronic and acute, neuralgia, with its accompaniment of partial paralysis and various other complications, torpor of the lower extremities of a child, nearly amounting to paralysis, have all yielded to the application of these vital forces of light. May we not congratulate mankind on the blessings which this discovery foreshadows?

For cerebral disorders, from softening of the brain to confirmed insanity, I would respectfully suggest to the medical profession full trials of the blue and sunlight baths, to be taken by their patients at least once in every twenty-four hours on the naked spine and back of the head. Should they succeed in removing the disorders of the brain, we may, in the near future, be relieved of the cost of building additional lunatic asylums, and insanity may be classed as a curable disease.

While this edition was being put through the press, I received the following communication and its enclosure from Dr. Robert Rohland, a distinguished scientist, resident in New York:

209 Third Avenue, New York.
October 26th, 1874.

Gen. A. J. Pleasonton.

Dear Sir:—With my warmest thanks for your last kind letter, I have, to-day, the pleasure to send you enclosed, at last, the report of Dr. Fisher's patient; and am still in hopes to send you more next month.

Accept the assurance of my highest respect, and allow me to sign myself, your most obedient and grateful,

Dr. ROBERT ROHLAND.

Enclosed in the above, was the following statement of the lady who had been placed under the influence of the associated light of the sun and the blue light of the firmament, and the blue rays eliminated from sun-light transmitted through blue glass:

"At the request of my attending physician, Dr. Louis Fisher, I will state, as briefly as possible, the effects produced upon me by the transmission of the sun's rays through blue glass:

"Having been an invalid for nearly three years, and for the last half of that time confined entirely to my rooms on one floor, I became so reduced by the long confinement, and my nervous system seemed so completely broken down, that all tonics lost their effects, sleep at nights could only be obtained by the use of opiates, appetite, of course, there was none, and scarcely a vestige of color remained, either in my lips, face or hands—as a last resort I was placed, about the 19th of January, 1874, under the influence of blue glass rays. Two large panes of the glass, each 36 inches long by 16 inches wide, were placed in the upper part of a sunny window in my parlour, a window with a south exposure, and as the blue and sunlight streamed into the room, I sat in it continuously—I was also advised by Dr. Fisher, to take a regular sun-bath of it; at least to let the blue rays fall directly on the spine for about 20 or 30 minutes at a time, morning and afternoon; but the effects of it were too strong for me to bear; and as I was progressing very favorably by merely sitting in it in my ordinary dress, that was considered sufficient.

"In two or three weeks the change began to be very perceptible. The colour began returning to my face, lips and hands, my nights became better, my appetite more natural, and my strength and vitality to return, while my whole nervous system, was most decidedly strengthened and soothed.

"In about six weeks, I was allowed to try going up and down a few stairs at a time, being able to test in that way how the strength was returning into my limbs, and by the middle of April, when the spring was sufficiently advanced to make it prudent for me to try walking out, I was able to do so.

"The experiment was made a peculiarly fair one by the stoppage of all tonics, &c., as soon as the glass was placed in the window, allowing me to depend solely on the efficacy of the blue light."

A distinguished surgeon of this city, on being made acquainted with the remarkable vivifying effects of this force, in several of the cases mentioned herein, expressed to the author, the opinion that the vitalizing influence of these associated colours, would probably be found to eradicate scrofula, and the terrible diseases which have produced it, from the human system—a result never yet attained by any medical treatment now known.

If this opinion should prove to be well founded, why may we not anticipate that tubercular consumption of the lungs may be arrested in its progress, its abscesses absorbed and dispersed by the purified blood taking up the purulent matter, and either decomposing it, or eliminating it through the various excreting channels of the body?*

If this last mentioned case had furnished the only example of the restorative influence of blue light upon disordered health, it should awaken in the medical profession, throughout the world, a desire to investigate the causes and sources of that force which had produced such marvelous effects.

Let us attempt a solution. The juxtaposition of plain uncoloured glass and blue glass in the passage of sunlight, and the transmitted blue light of the firmament, and the eliminated blue rays of the sun-light through them respectively, evolves an electro-magnetic current, which imparts to vegetable or animal life subjected to it, an extraordinary impulse to the developement of their respective vigour and growth. Their vitality is strengthened so as to resist disease, and to throw it off in those instances in which it had appeared before having been subjected to its power.

---

* A friend of mine has sent me the following notice, viz:

"Life Under Glass."—The author of "Life Under Glass," sends to the *Boston Transcript*, a letter giving some curious results of his experience in the use of coloured glass, as a medium for the transmission of the sun's rays in the treatment of lung disease. The writer of the communication, being himself a victim to weak lungs, gave special attention to the subject from personal as well as professional interest. His attention was directed to the matter by an accident in his own experience. During the autumn of 1863, he was home on "sick leave" from the army, and was in the habit of frequenting the photograph gallery of a friend. The operating room of the gallery was lighted by a skylight of light blue glass, and the walls were tinted of the same colour. He soon noticed, that he invariably felt better after an hour or two passed in the gallery, and he was firmly convinced that the beneficial effect was largely due to blue light. After the war, he began a series of experiments among his patients by using blue glass. As the light from pure blue glass is not entirely agreeable to the eye, he alternated the panes with clear glass. This was an improvement, and he went on with his experiment until he attained the highest sanitary power in a purple or light violet colour, the red, in the staining, making the light pleasant to bear.

The velocity of light on the earth's surface has been found by Leon Foucault, by experiments most carefully conducted, to be 298,000 kilometres or 186,000 miles per second of time—now of the seven primary rays of light, all of them excepting the blue ray and possibly its compounds, purple, indigo and violet, which perhaps are decomposed, and the blue ray liberated, are suddenly arrested in their marvelously rapid course, on coming in contract with the blue glass. This sudden impact of the intercepted rays on the outer surface of the blue glass with this inconceivable speed, produces a large amount of friction. Light, though imponderable, yet is material, since according to the book of Genesis, God said, "Let light be made, and it was made"—and the movement of matter upon matter, always produces friction. By friction electricity is evolved, and when opposite electricities meet in conjunction, their conflict according to the celebrated Danish philosopher, Oërsted, develops magnetism. The electricity produced by this friction is negative, while the electrical condition of the glass is opposite, or positive, and heat is therefore also evolved by their conjunction. This heat sufficiently expands the pores of the glass to pass through it—and then you have within the apartment, electricity, magnetism, light and heat—all essential elements of vital force. Without light and heat, life cannot exist, and electricity and magnetism are indispensable to its active vitality. This current of electro-magnetism, when allowed to fall upon the spinal column of an animal, is conducted by its nerves to the brain, and thence is distributed over its whole nervous system, imparting vigour to all the organs of the body, and stimulating them into active exercise: hence follows restoration to health.

In the early part of the summer of 1871, having caused to be printed an edition of my memoir which, a short time before, I had read before you, I distributed copies of it among literary and scientific institutions, and to such persons of culture as were likely to be interested in the investigation of the subjects treated of in it. Having sent several copies to Washington city, I received from my friends there suggestions to take out Letters Patent from the Government of the United States for my new discovery, which they deemed to be of the highest importance. Accordingly, I made an application to the Commissioner of Patents for the issue of Letters Patent thereon. When the application was received at the Patent Office, the novelty of its character, and the wonderful results of the experiments on which the application had been based, excited the greatest surprise and interest among the officers of the Bureau of Patents. The application was referred by the Commissioner to the Examiner-in-chief of the class of Chemistry, who, after a full examination of the whole subject, as I was informed, reported favourably upon the application and recommended the issue of Letters Patent. At this stage of the proceeding, the Commissioner was visited by the Examiner-in-chief of the class of Agriculture, Professor I. Brainerd, of Ohio, a very distinguished scientific gentleman, who suggested to the Commissioner that the application had received a wrong reference; that it should have been referred to him as it concerned plants and animals, which were intimately associated with the class of Agriculture under his charge. The Commissioner replied, that it concerned, also, Chemistry; but if he, Professor Brainerd, desired to investigate the subject, the issue of the Letters Patent should be suspended till that opportunity was afforded him—which was done. I was thereupon informed of it, and that the Commissioner, in view of the great importance of the application, and of the novelty of the principles involved in it, was desirous, before proceeding further in the issue of the Letters Patent, to send to my farm in this vicinity, Professor Brainerd, who, with my permission, would examine into the manner in which my experiments had been conducted, and particularly investigate the whole subject of the application. On the receipt of this communication, I wrote to the Commissioner of Patents, and informed him that I would be very glad to receive Professor Brainerd, and to give him every information and afford him every facility for making his investigation in my power.

A few days thereafter, the Professor arrived at my house in Spruce street; and, on presenting himself to me, he said: "General, you must receive this visit of mine as a very high compliment, since the Commissioner of Patents, in extremely rare cases, ever sends any one from the office for information in relation to an application for a Patent; for he requires all such information to be brought to him. He has, however, in this case deviated from his usual course, from the great interest he feels in your alleged discovery, and has sent me, therefore, to make the necessary investigation. For myself, I will say, that I have no prejudice for or against the principles announced in your startling memoir, and I come to you to make a fair, honest and impartial examination of the whole matter. If your averments, General, shall be sustained after I shall have examined the subject, I will report favourably upon your application, and your Letters Patent will be issued forthwith. Should I, however, have any doubts in the matter I will report against their issue, and you will not get your Patent." To this I replied, "That the facts in the case must furnish their own evidence, and I was perfectly satisfied to abide by his judgment thereon, whatever it might be." We then proceeded to my farm, where the professor remained three days, devoting himself to a critical examination of the subjects committed to him for investigation. On the afternoon of the third day we visited the grapery, as he had often done before, where we met three professors of colleges, who, attracted by the notices of the experiments which they had seen in the newspapers, had come to the farm to verify for themselves the statements the had read. For purposes of ventilation in the grapery, I had caused to be removed from immediately below the eaves on the southeastern side thereof, for the whole length of the house, two panes of glass in width; and in their places I had introduced galvanized iron wire cloth, with meshes of about one quarter of an inch square. The vines planted on the outside border, and trained through terra-cotta pipes into the grapery, along its walls of glass, and up to the ridge on the southeastern side of the grapery had, when they reached this wire cloth, in their growth on the inside, sent lateral branches through its meshes into the outer air, which had grown to varying lengths of ten, twelve or fourteen feet on the outside of the grapery. These lateral branches were covered with foliage—the inside branches from the same stems extending to the ridge were likewise covered with the densest foliage; but the difference between the inside and outside foliage was most distinctly marked. The inside leaves, from the same roots which furnished those on the outside, were fully six or eight inches respectively in diameter, of the deepest green colour, and so perfectly healthy that they seemed more like wax leaves than natural ones, while those on the outside of the grapery, though abundant, were not more than two inches in diameter, of a pale, sickly, yellowish colour, indicating a feeble vitality. I called the attention of Professor Brainerd and of the other professors to this most marked difference in the respective leaves inside and outside, and they all united in the opinion that this example furnished the most conclusive illustration of the influence of blue light on vegetation that could be produced under any circumstances. Here were branches of vines from the same roots, covered with foliage, deriving their nutriment from the same sources, the outside leaves exposed to all the influences of temperature, light, humidity or dryness of the natural atmosphere, and yet, scarcely one-fourth of the size of their relatives—those on the inside; and indicating an enfeebled and transitory existence. While the latter, revelling in the stimulating forces of the combined sunlight and blue light of the sky, had attained not merely size, but also an exuberance of vigor which excited the greatest astonishment. Professor Brainerd gathered some of the leaves from the outside and inside branches of the same vines, which he took with him to the Patent Office to be measured and photographed. The other professors did likewise to exhibit to their respective classes.

When Professor Brainerd had completed his examination, and was prepared to return to Washington, he said to me, "General, everything that you have alleged on this subject of blue light is confirmed; I am perfectly convinced of their truth. On my return to Washington, I will make a most favourable report on your application, and your Letters Patent will be issued forthwith. I will now say to you, that before I left Washington, the officers of the Patent Office discussed among ourselves your application, and we came to the conclusion, unanimously, that if my investigation should establish the verity of your statements you have made the most important discovery of this century, transcending in importance even that of Morse's Telegraph, which, at best, furnished only a means of communication with distant places, while your discovery could be brought home to every living object on the planet. We further thought that your patent would be one of the most valuable that had ever been issued in the United States. I congratulate you upon your great discovery."

The Professor accordingly returned to Washington, made his report, which, as he said it would be, was most favourable; and Letters Patent for my new process of accelerating the growth of plants and animals were issued to me on September 26th, 1871.

It is to Moses, the lawgiver, the great leader of the Israelites in their Exodus from Egypt, in their passage across the Red Sea, and in their subsequent residence in the desert, that we are indebted for our knowledge of the plan of the Deity in the creation of the world. This narrative of Moses, as contained in the book of Genesis, has been received by Christian and Jewish peoples, of all nations, as a faithful description of the revelations claimed by Moses to have been made to him by the Almighty himself. It is the foundation of their religions—the basis on which their spiritual faiths rest.

Let us take up this book of Genesis, and endeavour to discover from it, illuminated by the developments of modern science, what the prevailing idea of the creative mind may have been in establishing the physical functions of the planet on which we live.

In the first chapter of Genesis, we read the first four verses as follows, viz:

"1. In the beginning God created heaven and earth.

"2. And the earth was void and empty, and darkness was upon the face of the deep, and the Spirit of God moved on the waters.

"3. And God said, Be light made: and light was made.

"4. And God saw the light that it was good, and he divided the light from the darkness."

From these verses, it would appear that the materials composing this planet were created and assembled in darkness, and that the first physical force made was light—not heat, not electricity, not magnetism—but light, which we shall endeavour to show is the almost omnipotent force, which produces them all, and gives form and motion to our planetary system. In the same chapter, in the 6th verse, we read,

"6. And God said; Let there be a firmament made amidst the waters, and let it divide the waters from the waters."

And in the 7th verse, we read as follows, viz:

"7. And God made a firmament, and divided the waters that were under the firmament from those that were above the firmament—and it was so."

There is obscurity in this verse, since in the following verse, the 8th, we read,

"8. God called the firmament Heaven,—and the evening and the morning were the second day." Now in the 1st verse it is stated, "In the beginning God created heaven and earth;" heaven having precedence both as to time and place in the creation. In the 8th verse, it would read as if there were waters above the heaven, which were divided by the firmament from those that were on the earth. We may suppose, therefore, the word firmament, used in the 7th verse, to mean the atmosphere, which was to hold in suspension the waters contained in it as vapours, clouds, &c., thus separating them from the waters on the earth, as well as the infinite space above the atmosphere, now supposed to contain the orbits of the fixed stars. In the 9th verse, the dry land appears, and the waters under the heaven (probably atmosphere) are gathered together and, in the 10th verse, are called seas, and in the 11th verse God said, "11. Let the earth bring forth the green herb, and such as may seed, and the fruit tree yielding fruit after its kind which may have seed in itself upon the earth, and it was done.

"12. And the earth brought forth the green herb, and such as yieldeth seed according to its kind, and the tree that beareth fruit having seed, each one according to its kind, and God saw that it was good."

We will here observe, that so far as the order of developing creation had gone, light was, as yet, the only active force which had been brought into existence, or as the verse expressed it, "and light was made." Of course, it must have been made of the materials which composed it. There were, at that period, no sun, no moon, and perhaps only the fixed stars, which were to illuminate the heaven, that had been created, and yet light was made, and it was made of its materials, and being made its attributes were at once called into use. "For the earth brought forth the green herb, and such as yieldeth seed according to its kind, and the tree that beareth fruit having seed, each one according to its kind." No herb could have been green without light, and no tree could have borne its fruit in darkness, nor could seed have been matured without light, and yet this light came neither from the sun, nor the moon, modern spectroscopes to the contrary notwithstanding, for as yet neither the sun nor the moon had been created.

Hence, we can understand that the Creator, in directing that light first of all should be made, intended to constitute a force superior to all other forces, for it is by light that they are all developed, and made auxiliary to the great plan of Creation.

"14. And God said, Let there be lights made in the firmament of heaven, to divide the day and the night, and let them be for signs and seasons and for days and years.

"15. To shine in the firmament of heaven and to give light upon the earth, and it was so done.

"16. And God made two great lights, a greater light to rule the day, and a lesser light to rule the night, and the stars.

"17. And he set them in the firmament of heaven to shine upon the earth,

"18. And to rule the day and the night, and to divide the light and the darkness, and God saw that it was good."

It will be seen from these verses, that the ruling intent of the Creator was to furnish *light*, and not heat, to the world he was bringing into existence—to separate the day from the night—as signs and for seasons, and for days and years, to shine in the firmament of heaven, and to give *light* upon the earth.

These then are the varied functions to be performed by the sun, moon, and stars, by the fiat of the Creator.

Much speculation has been evoked, in the inquiry for the source of *that light* that was ordered to be made previous to the making of the two great lights, the sun and moon, which he set in the firmament of heaven to shine upon the earth. The modern revelations of the telescope in disclosing the character of the more distant fixed stars, the congregations of stars in the "Milky Way," in the nebulæ and cloudlets of lights, furnish an answer to all such inquiries. The limited vision of Moses, unassisted by the telescope, which, in his day, had no existence, would not have permitted him to comprehend any revelation of the glories of the world of astronomy, as known to us now; and hence, no such revelation was made to him. He was only instructed partially on the subject of our solar system, and the myriads of lights, lesser and greater than any that our system contains, which were sending their illumination over a boundless world, were entirely unimagined by him. But we can readily fancy with our increased knowledge of astronomy, whence this primeval light was drawn. We may suppose that our solar system was the last created of the various systems which stud the heavens with their brilliant effulgence, and that the materials which compose it were easily gathered from the mighty masses that illuminated the firmament.

Our astronomers tell us of the infinite star depths, in which are assembled series of worlds without number, all circling around their respective central orbs, and all moving with inconceivable velocity towards some region of the firmament so remote that our finite intellectual powers fail to conceive of it, and that, in this grand movement of worlds, our diminutive solar system has its allotted part and pursues its inevitable destiny. Hence arises the reflection that when our system shall approach the astronomical horizon of this mighty system of worlds, and shall be descending below it, as our sun now does below our own horizen, another solar system, transcending in its glories anything of which the human mind can conceive, shall arise in the western firmament to take the place that had been vacated by our own, and thus system after system shall be circling in the great expanse of space, till time shall be no more.

We must have a starting point in our discussion, and we will begin with matter, out of which all things are made.

We define matter to be anything which moves, or is the subject of motion. We prefer this definition before all others, since it is entirely irrespective of human existence, and has no reference to human impressions. Motion was produced long before man, and will continue long after he has passed away.

When matter is said to be solid, liquid or gaseous, we convey a very inadequate idea of its composition or of its condition. The microscope, as its powers are being developed, reveals to us forms and conditions of matter of which the most fertile imagination could have had no previous conception. So in the series of what is termed created matter, we have but a very faint image of a few of the most obvious links in the chain of its conditions, while we know and can know nothing of its extreme terminations, its greatest density and most minute tenuity. But we may conceive that whatever moves, or can be moved, must be matter—according to this definition, the imponderables, light, heat, electricity and magnetism, are all material substances, so subtle and attenuated, however, that human ingenuity has never been able to discover their components, or to reduce them to standards of comparison by which their powers might be measured. We might go farther and assert that all human emotions as well as animal instincts are likewise material; since our only cognizance of them is made apparent to us through our senses, concerning whose materiality there can be no question. Let it not be supposed that this idea of material being is at all inconsistent with an aspiration for a future life, since the resurrection of the material body is as much a part of the Christian's creed as is the hope of his immortality. Moses has told us for what purposes the sun and moon and stars were created; "to rule the day and night, and to divide the light and the darkness, and as signs, and for seasons, and for days and years." Now, it is a very remarkable thing, that Moses, who was born in Goshen, a province of Egypt, who passed the first forty years of his life in Egypt, which lies between north latitude 32° and 22°, and 27° and 34° east longitude, the next forty years on the borders of the Desert, and the last forty years thereof in the wilderness with his people, should have omitted to assign to the sun the heating qualities which our scientists declare it to possess, if, in fact, the sun did possess such powers, and the fact had been revealed to him by the Almighty.

Modern discoveries in science go to show that Moses was right in his description of the functions of those luminaries.

We may imagine the astonishment, amounting almost to incredulity, with which Moses received the revelation regarding the attributes of the sun, moon and stars. Living in the hot climate of Egypt, or of the Desert, whose "soil is fire, and whose wind is flame," and termed "burning sands of the Desert," from their great heat, to what other source could he refer this terrible heat than to the sun. Yet the sun is described to him as a *great light*, not a great furnace, not a great source of heat, but simply as an illuminating power. When traveling in the Desert, and overtaken by the burning Sirocco, whose blast, like that from a fiery furnace, obscuring the light of the sun by the clouds of burning sand which it had raised, Moses might have, by a course of reasoning, traced a connection between the raging tempest and the sands heated by the sun, and thus have assigned to that luminary the heating power claimed for its radiations. He might even have been familiar with the tenets of the predecessors of Zoroaster, and of the fire worshippers in Persia, who worshipped that great orb of light as the source of earthly heat, but if so, he discarded all such imaginings, and boldly declared "that it is the greater of two *lights*, intended to separate the day from the night; as signs, and for seasons, and for days and years; to shine in the firmament of Heaven, and to give *light* upon the earth."

Light is the great source of terrestrial electricity, magnetism and heat.

Whatever moves, or is the subject of motion, is matter.

We cannot conceive of motion, without associating with the idea an object to be moved. Hence light, which moves with a velocity of which we may speak, but which is not conceivable by us, is composed of matter. When the Creator, in his beneficence, first displayed the rainbow in the atmosphere, he taught mankind their first lesson in philosophical analysis. He thus showed that the white light of the sun was not a simple substance, but that it was composed of seven primary rays, which, by their combinations, produced all the varying tints or colours that are seen in nature, and yet how many myriads of years have passed since this magnificent spectacle has been exhibited to man before any one ventured to inquire into the simple and beautiful lesson which it taught. Even yet, what profound ignorance prevails everywhere in connection with the influences which these elementary rays develop.

Light, which thrown upon the photosphere of the sun, from the innumerable orbs that from their starry depths illuminate the expanse of Heaven, is reflected to this planet with a velocity of 186,000 miles per second of time, and requires about 8 16-35 minutes to reach the earth from the sun, ninety-two millions of miles distant. Whatever may be the composition of the space intervening between the sun and the earth, outside of our atmosphere, as we are taught that nature abhors a vacuum, it must be composed of something which is made of matter. Give it its most attenuated form and call it ether, it is still matter, and light, which is also composed of matter, however subtle it may be, passing through it with this marvelous speed, must produce everywhere enormous friction. Now whenever one body moves in, on, under, around, or through another body in contact with it, such motion produces friction. Friction, derived according to Webster, from the Latin *frico*, to rub, as we know evolves electricity, and it is this electricity and its correlative magnetism, discovered by Oersted, the celebrated Danish naturalist, to be its constant accompaniment when opposite electrical polarities are united, thus derived, which form those tremendous forces of nature that produce everywhere those changes in, on and about our planet, that meet our observation at every instant. When, therefore, the Creator, after having assembled in their respective positions the materials which compose the planetary and stellar worlds, uttered the sublime words, "Let Light be made," he called into being a power which became the generator of all the physical forces which control and regulate the world. Let us for a moment imagine the radiant reflection of luminous matter from every part of the photosphere of that great luminary, the sun, which in its magnitude was intended to illumine and vitalize all animated matter, as well as to give form and consistency to whatever had been created, passing from every point thereof with a velocity of 186,000 miles per second, penetrating through planetary and stellar spaces which, however subtle and attenuated, must have offered some resistance to the passage of this material light, producing everywhere in its passage an enormous amount of friction, and with it electricity and magnetism. Electricity, by the junction of its opposite polarities, evolves heat and also imparts to all substances that are capable of being invested with it, magnetism. The sun, the planets, the stars and all the bodies that stud the expanse of heaven, are doubtless all magnets, to which magnetism was imparted when the Creator uttered in heaven the words without parallel in sublimity, "Let light be made." This then is the origin of all the physical forces of the universe. Let us consider for a moment the nature of heat, and it will be apparent that terrestrial heat cannot be directly derived from the sun.

The tendency of heat is always to ascend into the atmosphere, when it is derived from combustion on the surface of the earth, or from radiation within it. The flame of a candle is vertically upward, on every part of the earth's surface, when the air is still. The effort of heat is to depart from its source with a rapidity proportionate to the intensity of the combustion. This is a repellent force—at the same time from its being associated with positive electricity, it is attracted to the upper atmosphere by its negative electricity, always associated with cold, which is opposed to positive electricity. The diffusion of heat, laterally or downwards, is very inconsiderable, as is constantly manifested in our rooms, where the fire in the grate emits very little heat below the bottom of the grate, and parts of the room distant from the fire are very imperfectly heated by it. The sun in its daily course being above the earth, if it had any calorific rays, could not send them to the earth below it, through a space of ninety-two millions of miles, which, according to calculations of Pouillet, has a temperature of minus 142 degrees of Centigrade thermometer. We will illustrate this by an example or two. During our late unhappy sectional war, at the siege of Fort Sumter, in South Carolina, General Gilmore's heavy guns threw their enormous shells into the city of Charleston, four and a half miles distant. While the expansion of the powder in the chamber of these guns, in its combustion into gases, evolved a power which threw these shells so great a distance, it was totally inadequate to drive the heat disengaged in the conversion of the powder into these propelling gases to a greater distance from the muzzles of the guns than thirty feet. It ascended, instantly on leaving the guns, into the upper atmosphere, attracted by an opposite electricity. Any one familiar with the fire of artillery, must have observed similar effects regarding the heat from the discharge.

We will illustrate this by an example. "Mount Washington, in the White Mountains, in New Hampshire, is in north latitude 44° 16′ 25″, and in west longitude from Greenwich 71° 16′ 26″. Its elevation above tide water is 6,293 feet; and in altitude it is the second highest mountain northward of the Gulf of Mexico, the highest mountain thereof being Clingmans Peak, in the State of North Carolina—which is 6,707 feet above tide water.

"The limit of the growth of trees on the north side of Mount Washington is 4,150 feet above tide water. The climate of Mount Washington corresponds with that of the middle of Greenland, about 70° of north latitude or 26° further north than New Hampshire. It is an arctic island (so to speak) in the temperate zone, and, on account of its great elevation, it exhibits also the condition of the atmosphere where the mercury does not rise above 24 inches in the barometer. For peculiar interest, therefore, the Mount Washington (meteorological) station is not exceeded by any point within the arctic circle."

It was on this mountain that a party of scientific gentlemen passed the winter of 1870 and 1871, amid great privations and suffering, for the purpose of investigating the physical conditions of the atmosphere and mountain at that great elevation. "Observation shows that the climate of any country becomes colder in proportion to the height of the land above the sea. Thus in tropical regions there may be an arctic climate at an altitude of 12,000 or 15,000 feet."

The room inhabited by these gentlemen was in the south-west corner of the railroad depot, about 20 feet long, 11 feet wide and 8 feet high. It was well protected from the outer cold, was heated by two stoves, one an ordinary cook stove, the other a Magee parlor stove, prized for its marvelous heating power. Their Journal reports as follows, viz:

"February 4th, 1871, temperature at 7 o'clock, A. M., —33°; at 9 o'clock, P. M., —40°. In the room the temperature was +35° and sometimes +60°. To do this, the stoves were kept at a red heat. The thermometer hangs 5 feet from stoves, the temperature 10 feet from the stoves at the floor was 12°, in other parts of the room the temperature was 65°; midnight, wind fully up to 100 miles per hour and northwest.

"February 5th, some of the gusts of wind 110 miles per hour; at 3 o'clock, A. M., temperature in the room 59°, barometer 22.810 inches, attached thermometer 62°. Yesterday, barometer 22.508 inches."

Now let us see what this means: 5 feet from red hot stoves the thermometer marked 60°, 10 feet from the same stoves on the floor the thermometer marked 12°, being a loss of 48° in a distance of 5 feet in length and 2 feet below the sources of heat. Now at that rate of radiation of heat, how hot must the sun be to transmit any degree of heat 92 millions of miles through a temperature of —142° of centigrade to this planet, and not merely to this earth in a column of heat of 8,000 miles in diameter to envelope it, but also to diffuse its heat through an ellipsoid of ether, whose circumference would be the orbit of the earth around the sun? But the actual loss of heat in its descent to the earth (if that could be possible, which it cannot be,) per foot would be immensely more than is stated above, as the heat would have to pass through space chilled to —142° of centigrade instead of in a room heated to +65° of Fahrenheit. Again, in this latitude of 40° north, we have in our winters falls of snow which lie upon the ground sometimes for weeks, with the sun being unable to make any impression upon it—and when the snow does begin to melt, it commences with the layer of snow in contact with the earth, and not with that on the upper surface exposed to the sun. Our farmers all know that when their fields in winter are covered with snow, their growing crops under it are kept warm, though no ray of the sun could reach them through the snow, and they anticipate therefrom a large yield in the ensuing harvest. If terrestrial heat is derived directly from the sun, how is this fact explained? A gentleman in the State of Maine, during the early part of the last winter, when the ground at his residence was deeply covered with snow in many places, made some experiments to ascertain the temperature of the earth under the snow. He found that the heat increased at the surface of the earth with the depth of the snow above it. The following is the account, viz:

Experiments were made in the winter of 1872–73, with a view to ascertain how far the soil is protected from cold by snow. For four successive days in winter, there being four inches in depth of snow on the ground on a level, the average temperature, immediately above the snow, was found to be fourteen degrees of Fahrenheit's thermometer below zero; immediately beneath the snow in contact with the earth, it was ten degrees above zero; being an increase of twenty-four degrees of temperature, occasioned by a covering of the earth with four inches of snow; and under a drift of snow two feet deep the temperature was twenty-seven degrees above zero; making an increase of temperature at the earth's surface under two feet of snow, of forty-one degrees of Fahrenheit over the temperature of the air just above the upper surface of the snow. No one can pretend that these variations of temperature were derived from the sun. Let us attempt an explanation of this phenomenon.

It is this. The radiation of heat from the interior of the earth, positively electrified, meeting at the surface of the earth with the snow in contact with it, negatively electrified, the conjunction of these opposite polarities of electricity evolves heat, melting the under layer of the snow, irrigating the plants under it with water moderately warm, and keeping the earth from being frozen, so that in the spring following, when the snow had disappeared, the plants were ready to receive the stimulating influence of sunlight and the blue light of the sky, of which they had been deprived during the winter.

Professor Tyndall, writing of what he calls solar radiation, says: "Never did I suffer so much from solar heat, as when descending from the *corridor* to the *grand plateau* of Mont Blanc on the 13th of August, 1857. Whilst I sank up to the waist in the snow, the sun darted its rays upon me with intolerable fierceness. On entering into the shade of the *Dôme du Goutè*, these impressions instantly changed, for the air was as cold as ice. It was not really much colder than the air traversed by the solar rays, and I suffered not from contact with warm air but from the stroke of the sun's rays, which reached me after passing through a medium as cold as ice."

It is singular that to so learned and astute a scientist as Professor Tyndall, it did not occur that if his sensations, so distressing on this occasion, were derived from the *heat* of the sun's fierce rays, that he could not have walked through snow waist deep, in such heat, without the snow becoming melted by the same heat which oppressed him, and that he would have been swept away by the torrent of water thus produced by the melting of the snow by this great heat; but it does not appear that the snow was at all affected by it, while the water was drawn out of the Professor in profuse perspiration.

I venture upon an explanation. The heat from which the Professor suffered came from his own body, and was derived from electrical action of sunlight upon his dark woolen clothes, warmed by the animal heat of his system. He was struggling through deep snow in an atmosphere of icy coldness. The natural heat of his body, ninety-eight degrees of temperature of Fahrenheit, was greatly increased by the muscular efforts he was making in his descent of the glacier. His woolen clothes had become positively electrified by the heat of his body. The strong sunlight of August having passed through the cold, dry ether of planetary space and the upper atmosphere of the earth, by its friction with them was negatively electrified, and falling upon his warm body and clothes, positively electrified, increased heat was evolved in and around his person, and his sufferings were intensified. As soon as he left the sunlight, his clothes, by induction, became negatively electrified and the temperature of his body was soon lowered, and his sufferings from heat ceased.

Again, there is no heat in the moon, which proves that the moon has not an atmosphere, as it also proves that there is no heat in the sun; for if there was an atmosphere about the moon the sun's light penetrating it and producing friction by the contact with it would evolve electricity, which uniting with the opposite electricity of the moon's atmosphere would produce heat, but no such effect has been perceptible with the most delicate instruments. Besides, if there was heat in the rays of the sunlight, that heat would be reflected with that light from the moon's surface to the earth, which we know is not the case.

Now, if the sun possessed heat, and could force it downwards to the earth, which, according to our knowledge of the laws of heat, is impossible, we could have no clouds in our atmosphere, as from the absorbing power of gases of heat the clouds would be so expanded and attenuated by the absorbed heat that they never could be formed.

The sun is a great magnet, as are all the planets of the solar system, and it is by their magnetism and not by their weight or gravitation that their motions in their respective orbits are regulated by the greater magnetism of the sun. Now as magnetic attraction or repulsion varies inversely as the squares of the distances, which relation has been heretofore attributed to gravitation, it is not difficult to assign to magnetism, in its attraction and repulsion, the forces which have heretofore kept and now keep our solar system in its various motions, nor need we hesitate to conceive that all the motions of infinite systems, of suns and stars, of nebulæ, and cometary and meteoric matter, are in like manner regulated. The meteoric matter which has fallen to the earth, has been found, when examined, to be highly magnetic.

If the sun is a magnet, there is only sufficient heat generated in its interior by opposite electricities to cause its daily rotation on its axis, and it cannot be an incandescent body, since magnetism is destroyed by heat.

Wherever there are differences of temperature, there are opposite electricities—one electricity being always associated with what is called heat while the opposite electricity accompanies cold. These terms of heat and cold are mere expressions of relative differences in varied temperatures, without regard to the intensity of either condition.

Professor Tyndall, in his book on "The Forms of Water in Clouds and Rivers, Ice and Glaciers," has given what he considers explanations of many physical phenomena connected with his subjects, attributing to radiations of solar heat the changes and transformations which he describes. With great deference to so learned and distinguished an authority, I take occasion to offer other explanations of the causes of the phenomena alluded to, which seem to me as being more in accordance with our knowledge of general physics.

In his article on "Mountain Condensers," he says: "Imagine a southwest wind blowing across the Atlantic towards Ireland. In its passage it charges itself with aqueous vapour. In the south of Ireland it encounters the mountains of Kerry; the highest of these is Magillicuddy's Reeks, near Killarney. Now the lowest stratum of this Atlantic wind is that which is most fully charged with vapour. When it encounters the base of the Kerry Mountains, it is tilted up and flows bodily over them. Its load of vapour is therefore carried to a height, it expands on reaching the height, it is chilled in consequence of the expansion, and comes down in copious showers of rain. From this, in fact, arises the luxuriant vegetation of Killarney; to this indeed, the lakes owe their water supply. The cold crests of the mountain also aid in the work of condensation."

Let us examine this. The tilting up of the masses of cloud on coming in contact with the face of the mountain is the resultant of the impact of two forces, one being that of the wind from the southwest with any given velocity from twenty miles per hour to that of eighty or one hundred miles per hour; the other, the static force of the resistance of the mountain itself; the diagonal of these two forces is the tilting up of the cloud after impact. Now these two great masses of cloud and mountain, oppositely electrified, when they come together in contact produce great friction of their molecules, which friction evolves positive electricity from the higher temperature of the southwest wind; this positive electricity thus evolved rushes into conjunction with the opposite electricity of the atmosphere, producing heat, which heat being absorbed by the air holding the water in suspension communicates to it positive electricity, and the air so electrified is attracted by the negative electricity of the upper atmosphere, carrying it up and by expansion so comminuting the particles of air that they can no longer contain the globules of water they before held in suspension, which latter thus released then begin, being attracted by the positive electricity of the earth, to fall as rain oppositely electrified, and it is, therefore, these electricities thus excited with the heat which is evolved by their conjunction and the rain charged with ammonia and carbonic acid gas which furnish the stimulants to the remarkable vegetation of Killarney. During the prevalence of these rain bearing clouds, driven across the Atlantic by the southwest winds upon the above mentioned mountains, the sun must be obscured by them, and hence there can be no radiations of solar heat to expand the air of the clouds after their impact with the mountains, and they have been tilted up in their further progress over the crests of the mountains.

A similar explanation covers the example the Professor gives of a heavy fall of rain or snow in the Alps, while the sky is clear and blue over the plains of Italy—*while the wind is blowing over the plains to the Alps.* The warm wind, positively electrified and holding water in suspension, coming in contact with the negative electricity of the cold Alps, and producing friction by the impact, evolving more positive electricity to combine with the negative electricity of the atmosphere at that great elevation, increases the heat, and by it expands the air of the clouds so much that it can no longer hold the globules of water held by it in suspension. The heated and expanded air, attracted to the still higher atmosphere from its greater negative electricity, separates from the water it before held, while the water having lost its heat by the superior capacity of the air to absorb it, becomes negatively electrified and is attracted to the earth by its positive electricity—hence the rain fall.

Professor Tyndall also states in the same work, "that the unconfined air heated at the earth's surface, and ascending by its lightness, must expand more and more, the higher it rises in the atmosphere," and that *the ascending* "air is chilled by its expansion. Indeed this chilling is one source of the coldness of the higher atmospheric regions." It strikes me that this explanation is not correct. In the first place the ascent of heated air in the upper atmosphere has a limit beyond which it cannot pass. Secondly, it ascends not by its lightness but by the attraction of the negative electricity of the upper atmosphere for the heated air, which is oppositely electrified. In its upward course it loses its heat by radiation and with it its positive electricity—and by induction becomes negatively electrified with the air whose altitude it has reached—nor is this chilling by expansion, as he terms it, one source of the coldness of the upper atmosphere. That coldness associated with negative electricity is derived from the ether in which the atmosphere as well as the earth is continually revolving; that ether has a temperature, according to Pouillet, cf —142° of Centigrade thermometer, and our upper atmosphere in contact with this ether receives from it, by induction, both its cold and its negative electricity, and the atmosphere itself is kept in its place as an envelope of the earth by the positive electricity of the earth and the opposite electricity of the upper atmosphere. The snow line from the equator, (15,000 feet above the equator to the 60° of north latitude, where it coincides with the earth,) being the dividing line between these two opposing electricities.

The Professor gives another example of the air being chilled by its expansion, as follows, viz: "with a condensing syringe you can force air into an iron box furnished with a stop cock, to which the syringe is screwed. Do so till the density of the air within the box is doubled or trebled. Immediately after this condensation, both the box and the air within it are warm, and can be proved to be so by a proper thermometer. Simply turn the cock and allow the compressed air to stream into the atmosphere. The current, if allowed to strike a thermometer visibly chills it, even the hands feel the chill of the expanding air."

Now for another explanation different from the Professor's. The air in the iron box had become heated by the friction of it with the sides of the box; that friction evolved positive electricity associated with the heat; on turning the cock and allowing the heated air to escape into the atmosphere, the heat and the positive electricity both left the escaping air with the velocity of lightning, rushing into the oppositely electrified air in the upper atmosphere, and the air that reached the thermometer deprived of its heat reduced its temperature. There is also an inconsistency in the explanation of the Professor in producing heat by condensation in his iron box, while he produces rain by the condensation of the clouds by cold in the upper atmosphere. This reminds one of the fable of Æsop, in which a satyr invited into a husbandman's hut, blew upon his hot broth as he said to cool it before eating it, and again blew his breath upon his fingers to warm them on coming into the house from the cold outside air. The husbandman turned the satyr out of doors, as he could not comprehend how any one could blow hot and cold from the same breath.

If compression of the atmosphere produces heat, condensation, which is merely another form of expression for the same thing, cannot produce cold. If cold condenses, why does it not condense the air in the upper atmosphere where the greatest cold prevails, and the air is very dry, rarefied and attenuated? According to the theory of condensation by cold, the air should be very much more dense at great elevations above the earth, than it is at the surface of the ocean, but the reverse is known to be the case. The higher in the atmosphere a balloon, inflated with hydrogen gas, ascends, the more the gas becomes expanded by the rarefaction of the atmosphere, which shows that the cold of the upper atmosphere cannot condense the gas in opposition to the expansive influence of the rarefied atmosphere at great elevations. Ice water poured into a glass tumbler in the heat of summer, causes a deposit of drops of water on the outside of the tumbler resembling dew, which is the result of a conjunction of opposite electricities, the glass and the air within and around it being warm and positively electrified, while the ice water is negatively electrified. Their conjunction evolves heat, which is absorbed by the molecules of air, holding in suspension the humidity of the atmosphere; these molecules, so heated, ascend immediately with inconceivable rapidity into the upper atmosphere, attracted by its opposite negative electricity, while the globules of water thus released from their suspension in the air on the outside of the glass, being now negatively electrified, are attracted by the vitreous or positive electricity of the glass tumbler and are deposited on it.

On the thirty-first day of March, A. D., 1872, I visited my farm to give directions to apply heat to start the growth of the vines in my grapery, at the commencement of the season. The weather was very cold, patches of ice and snow lay in places on the fields, which the sun, shining with great brilliancy through a remarkably clear atmosphere, was unable to soften or melt. No semblance of cloud or vapour was anywhere visible. In the open air, protected from sunlight, the thermometer (Fahrenheit's) marked 34 degrees, two degrees above the freezing point of water. On entering the grapery, in which there had been no artificial heat from fuel of any kind for the space of nearly a year, my son and myself were astonished at the great heat that there was within it. On examining the thermometer which hung on one of the middle posts of the grapery, completely sheltered from the sunlight, about four feet from the floor, we were amazed to find that it marked one hundred and ten degrees of Fahrenheit. Here was an increase of seventy-six degrees of temperature over that of the outside air, and produced by a film of glass not exceeding one-sixteenth of an inch in thickness, but associated as blue and plain glass. This extraordinary increase of temperature, manifested the supreme wisdom of the Creator in kindling this heat at the surface of the earth, where it was needed, by rays of light passing through a denser medium than air, instead of sending heat from the sun through ninety-two millions of miles of ether at a temperature of —142 degrees of Centigrade thermometer, in the passage through which so much of the said heat would have been lost by radiation.

I have had many occasions to observe since that date, that during the passage of strong sunlight through the blue and plain glass of the grapery, the temperature through the day, within the grapery, varied from one hundred degrees to one hundred and fifteen degrees, while that without, according to the seasons of the year, at the same times of the day would range from thirty-two degrees upward to sixty degrees or sixty-five degrees.

During the winter of 1871 and 1872, which, in this city, was a very cold and rigourous one, two ladies of my family residing on the northern side of Spruce streets east of Broad street, in this city, who, at my suggestion, had caused blue glass to be placed in one of the windows of their dwelling, associated with plain glass, informed me that they had observed that when the sun shone through those associated glasses in their window, the temperature of the room, though in mid-winter, was so much increased that on many occasions they had been obliged during sunlight to dispense entirely with the fire which, ordinarily, they kept in their room, or when the fire was suffered to remain, they found it necessary to lower the upper sashes of their windows, which were without the blue glass, in order to moderate the oppressive heat.

These examples go to illustrate the remark of a distinguished German scientist, made to a friend of mine after he had read an account of my experiments with blue light on animal and vegetable life. He said, "that the discovery of this extraordinary influence was destined to produce the most important and beneficial results on the comfort and happiness of mankind, throughout the civilized world. That fuel was everywhere recognized as one of the most indispensable elements of social and domestic economy. That it is, particularly in Europe, very expensive from its scarcity, which is becoming greater every year with its annual consumption, and in the northern parts of Europe, furs, skins of animals and the down of aquatic birds are extensively worn, sometimes with two or three suits at once of clothing, in order to preserve the animal heat of the body, owing to the great costliness of fuel and the severity of the cold.

"That even in England, apprehensions are being expressed of an exhaustion of their coal mines in the not distant future. Now since this wonderful discovery of General Pleasonton, of the influence of the blue light of the sky in developing animal and vegetable life, which is largely due to the heat and electricity developed by the passage of sunlight through these associated blue and plain glasses, I am of the opinion that during sunshine, for many hours in the day, by means of blue and colourless glass arranged together in doors and windows exposed to the sun, sufficient heat can be evolved to enable families, and work people in factories, to dispense with a large proportion of the fuel that they have heretofore been obliged to use. Let us say that one-half of the fuel heretofore required, can be saved by thus utilizing sunlight, and you will begin to comprehend how vast will be the benefit derived to mankind in the economy of fuel alone, by this discovery of General Pleasonton."

I have said that while the rays of the sun's light were one of the causes of terrestrial heat, yet there is no heat in them. This can be proved by any one, in the following experiment, viz: During winter, when the ground is covered with snow, and the temperature of the open air is at zero of Fahrenheit's thermometer, it will be found that the sun, however brightly shining, cannot melt the snow or ice on which it may shine. Take now a piece of black or brown silken or woolen cloth of any form and of convenient size, and place it on the snow in the shade where the sun does not reach it with his rays. The snow will not be melted under this cloth, which will have the same temperature as the snow; hence it is obvious that there is no heat either in the sunlight which could not melt the snow, nor in the coloured cloth whose temperature was the same as the shaded snow on which it had been placed; now take up the cloth, and place it on the snow where the sun can shine upon it. Let us observe the effect of this new position; the rays of the sun moving with a velocity of 186,000 miles per second are suddenly arrested by this cloth, which they cannot penetrate. This sudden stoppage of velocity produces friction, by the impact of the rays of light upon the cloth; electricity is evolved by the friction, having a polarity opposed to that of the cloth; instantly these opposite electricities rush together, producing heat, warming the cloth and melting the snow immediately under the cloth, by which the cloth begins to sink below the level of the snow, and if it shall be allowed to remain, it will melt the snow under it till the cloth shall rest upon the ground beneath, clear of the snow, and the surrounding snow shall enclose the cloth, of its exact size and form.

From this experiment, we conclude that the heat which melted the snow under the cloth was not derived from the sun as heat, but that the electricity produced by the impact of the sun's rays with the cloth oppositely electrified, through friction, evolved the heat which melted the snow.

Now suppose that instead of a single piece of this cloth having been placed upon the snow, you have put a series of pieces of the same cloth upon the snow. The same principle applies but a different action is observed. The cloth is a bad conductor of heat as well as of electricity, consequently the heat evolved by the conjunction of the opposite electricities produced by the friction of the rays of sunlight by impact on the cloth with the opposite electricity of the cloth, cannot descend through the cloth to any depth, being contrary to the laws of heat, but it immediately ascends into the atmosphere and escapes, while the edges of the series of pieces of cloth in contact with the snow become warmed by the conjunction of the opposite electricities, produced by the friction of the rays of light with the edges of the cloth and the cloth's electricity, and soon melt the snow in contact with them, till the pieces of cloth are left high and dry above the snow which surrounds them.

Glaciers—their Origin, Position, Duration, Changes and Movements.—Much has been written on these subjects, and many distinguished scientists have been greatly exercised to give a satisfactory explanation of the phenomena they have witnessed in connection with them.

It seems to me that glaciers are formed in the regions of perpetual snow by the deposition of snow in the valleys of the lofty mountains where they exist; clouds laden with vapour when they reach the neighbourhood of the mountains whose valleys are filled with glaciers, being positively electrified, encounter the negative electricity of the higher atmosphere. These opposite electricities meet in conjunction, heat is evolved—the air associated with water as vapour in the clouds being thus heated, is rarefied and expanded to such an extent that it can no longer retain its water, (while it ascends rapidly into the upper atmosphere attracted by its negative electricity,) which on being liberated from the air that held it as vapour is converted by the surrounding low temperature of its great altitude into flakes of snow, which having an opposite magnetism to the earth are attracted downward to it, and are at the same time repelled from the height where they are formed by the opposite magnetism prevailing there. The crystallization of these snow flakes is made in a vacuum, produced by the escape of this heated and rarefied air, and by absorbing the magnetism which is developed by the conjunction of the opposite electricities of the clouds and the atmosphere as they come together in contact, these magnetic snow flakes transfer it to the earth to replace the magnetism which is constantly leaving the earth in evaporations to escape into the upper atmosphere.

This, then, in all probability, is the origin of glaciers. The successive snow falls in the upper valleys of these elevated regions, by their magnetic attraction to the earth, serve to pack the snow, and to compress the lower portions of it into ice of greater or less density, according to its elevation in the atmosphere and the depth of the valleys in which the glaciers are formed. The effect, therefore, is that the bottom of the glacier is ice, while the upper part of it is snow, termed névé.

Crevasses are fissures of various depths and widths in the glacier, whose formation Professor Tyndall attributes to the effect of the *solar radiation of heat* upon the glaciers. He says, in his book on "The Forms of Water," &c., page 100, "first, then, you are to know that the *air* of our atmosphere is hardly heated at all by the rays of the sun, whether visible or invisible; the air is highly transparent to all kinds of rays, and it is only the scanty fraction to which it is not transparent that expend their force in warming it."

I have shown that heat ascends in our atmosphere by the attraction of the positive electricity with which it is always associated, by the negative electricity of the colder air in the upper regions of the atmosphere, and by its repulsion from the earth by its positive electricity; it is, therefore, contrary to the laws of heat that the sun should, can or could transmit *rays of heat* downward to this planet, and as these heat rays can not be so transmitted, they are therefore not present to be absorbed by the snow of the glacier or on the mountains. On page 98 of the same book, he says: "we have wrapped up our chain and are turning homewards after a hard day's work upon the Glacier du Géant, when under our feet, as if coming from the body of the glacier, an explosion is heard. Somewhat startled, we look inquiringly over the ice. The sound is repeated, several shots being fired in quick succession. They seem sometimes to our right, sometimes to our left, giving the impression that the glacier is breaking up, still nothing is to be seen.

"We closely scan the ice, and after an hour's strict search we discover the cause of the reports. They announce the birth of a crevasse. Through a pool upon the glacier, we notice air bubbles ascending, and find the bottom of the pool crossed by a narrow crack, from which the bubbles issue. Right and left from this pool, we trace the young fissure through long distances. It is sometimes almost too feeble to be seen, and at no place is it wide enough to admit a knife blade.

"It is difficult to believe that the formidable fissures, among which you and I have so often trodden with awe, should commence in this small way. Such, however, is the case. The great and gaping chasms on and above the icefalls of the Géant and the Talèfre begin as narrow cracks, which open gradually to crevasses. The crevasses are grandest on the higher névés, where they sometimes appear as long yawning fissures, and sometimes as chasms of irregular outline; *delicate blue light* shimmers from them, but this is gradually lost in the darkness of their profounder portions.

"Over the edges of the chasms, and mostly over the southern edges, hang a coping of snow, and from this depend like stalactites, rows of transparent icicles, ten, twenty, thirty feet long. These pendent spears constitute one of the most beautiful features of the higher crevasses. How are they produced? Evidently by the thawing of the snow. But why, when once thawed, should the water freeze again to solid spears?" Now let us examine this: if the supposed heat of the sun's rays, could melt the snow at the southern edges of the crevasse, why did not similar rays from the sun, conveying the like temperature, melt the general surface of the glacier, and produce thereby large pools of water on the surface of the glacier? Particularly, as the Professor states, "that the snow on which the sunbeams fall, absorbs the solar heat, and on a sunny day, you may see the summits of the high Alps glistening with the water of liquefaction. The *air* above, and around the mountain's may, at the same time, be many degrees below the freezing point in temperature."

If the surface of the snow on the mountains was melted by the solar heat, as the Professor supposes, what was there to arrest the streams of water thus produced, and to prevent them from flowing into the valleys occupied by the glaciers, and converting the glaciers themselves into mountain torrents, while at the same time the mountains were being denuded of snow? But we know that such results have not been produced. Above the snow line the mountains are perpetually covered with snow, and the glaciers have remained from a remote antiquity to attest that the snow does not absorb the heat of the sunbeams, for the simple reason that the sunbeams in themselves do not bring any heat from the sun to this planet.

In my early boyhood, I dwelt on the banks of the Potomac, a river fancifully named by the Indians, before the advent of the white man, "the river of swans," from the abundance of that water fowl that frequented its waters. Well do I remember, lying awake on the eve of our several winter holidays, when the river was deeply frozen, anticipating a day of splendid skating on the morrow, to have been often startled by the noise of a great explosion of the ice on the river, occasioned by the compression of the air beneath the ice, as the tide rising rapidly forced it upwards between the water and the ice, till its accumulation and compression would overcome the resistance of the ice, and a fissure would be opened in it extending sometimes for miles, and liberating the pent up air into the atmosphere. If the temperature of the night air was below the freezing point of water, as the tide receded the water which had filled the fissure, when the tide was full, was frozen into ice, and the track of the fissure could be marked on the next day by the film of thin ice that had been formed in it, as the tide was receding the night before.

In this way, air holes, so dangerous to travelers and skaters on the ice, are constantly formed on our rivers and streams, subject to the flow of the tides, and in lakes and mountain streams, they are also formed by the currents of water flowing downwards in a similar manner. In my later youth, I had observed similar effects from similar causes, produced on the ice of the river Hudson, at West Point. In short, fissures on the surface of anything, whether on the surface of the earth by volcanic eruptions in which lava, rocks, scoriæ, mud, boiling water, are thrown out from the interior, or by Geysers spouting their hot streams into the atmosphere, or the cracks in the ground produced by long continued droughts, evaporating the moisture contained in the soil, and even eruptive diseases among mankind or other animals whether wild or domestic, are all the results of interior forces, acting from the interior to their respective surfaces.

Now let us explain the crevasse on the glacier. The snow falls carry to the glacier large quantities of atmospheric air, which are confined between the glacier and the snow as it falls; every fall of snow presses its predecessors and the air they contain closer together against the ice. filling its vacancies with air. This column of air, thus pressed down upon and into the ice, encounters the air which has been enclosed between the bottom of the glacier and the earth on which the glacier rests,—this last mentioned air has been warmed by the radiation of heat from the interior of the earth, and has become positively electrified by it—the contact of this positively electrified air with the negatively electrified ice of the bottom of the glacier, evolves more heat, which, melting the lower stratum of ice of the glacier, constitutes the source of the stream of water that flows from the glacier. Such is the origin of the river Rhone.

This warm air, in its effort to rise through the glacier into the upper atmosphere negatively electrified, meets in the crevices everywhere abounding in the ice of the glacier, the air which has been forced down by the snow falls, and which last air is negatively electrified; the conjunction of these two airs oppositely electrified evolves heat, which expanding the air, displaces the ice of the glacier, forming channels for its escape into the upper atmosphere, and when it reaches the upper surface of the glacier, forces its way through it into the atmosphere in that minute fissure, which Professor Tyndall had such difficulty to discover. Again, this warm air as it escaped into the atmosphere, melted the edges of the ice or snow at the surface through which it passed, and through which it was visible in the air bubbles Professor T. described.

The melting of the lower stratum of ice of the glacier in contact with the earth produced by the heat evolved by the conjunction of the positive electricity of the earth with the negative electricity of the ice, is the cause of the subsidence of the body of the glacier, and the declivity of the valley itself is the cause of the glacier moving bodily downward in it. The fractures, strains, torsions of certain portions of the glacier are the results of the forces of expansion and contraction in the interior of the glacier, produced by variations of its interior temperature as mentioned above.

In this country, the winter of the years 1874 and 1875 has been an exceptional one. The cold has been of long, and almost uninterrupted continuance, and of great severity. The rivers in the Middle and Eastern States have been closed with ice, which has been of great density and depth, extending in some of their courses through the mountains even to the beds of their streams. The frozen condition of the waters has remained till late in the spring season; and from the accumulation of immense masses of ice in certain portions of these rivers, forming what were called ice-gorges, filling their entire width for the distance of miles in length, the most serious apprehensions were entertained of extraordinary damages to towns and villages, railways and canals, in the valleys of these rivers, that would be sustained by the sudden breaking up of these gorges of ice from rain-storms, and the melting of the snows on the mountains, which would produce the most extensive and alarming inundations. These apprehensions were justified by the advanced spring season which usually, by its increased temperature, terminates the rigours of winter.

To obviate, if possible, these threatened dangers and calamities by the sudden breaking up of the ice, various expedients were resorted to, viz: cutting channels through the ice below the gorges, to liberate the water above, should it assume alarming proportions; attempting to destroy the gorges themselves by the explosions of gunpowder, or of nitro-glycerine, confined in chambers in the ice itself, and one very liberal gentleman, evidently a believer in the theory that the sun is an incandescent body and sends its heat bodily to our earth, downwards, presented to the authorities of one of the towns endangered by the ice-gorge in its neighbourhood, twenty-eight barrels of Naphtha, to be burnt on the ice-gorge, under the impression that the heat produced by their combustion, would descend through the ice, and liquefy it into water. It is scarcely necessary to add, that, when the experiment of burning the Naphtha upon the ice-gorge was tried, the heat evolved by its combustion immediately ascended into the upper atmosphere, leaving the ice unaffected by the experiment.

From a very interesting book entitled, " Mount Washington in Winter; or, the Experiences of a Scientific Expedition upon the Highest Mountain in New England—1870–71," published in Boston in 1871, we make some extracts that seem to have a connection with the subjects of which we are treating.

" Moosilauke Mountain, near Mount Washington, is nearly five thousand feet high, and lies within the arctic zone of climate. It was on this mountain that two scientific gentlemen, viz., Messrs. A. F. Clough and H. A. Kimball, determined to pass two months, in the winter of the years 1869 and 1870, in order to fit themselves the better for a winter residence on Mount Washington, at a future day. They attempted the ascent of the mountain on November 23d, 1869, but were driven back by the severity of the weather. On the 31st of December, 1869, the attempt was renewed under better auspices, and was successful.

"About two months were spent by them on this summit. So valuable were the experiences acquired, and so unusual were the meteorological phenomena observed, that the Mount Washington phenomena, subsequently experienced, have not equaled those upon Mount Moosilauke, and among them the possibility of living on a mountain top during the winter, was fully demonstrated.

"There is scarcely a mountain in New England from which the view is more extensive. We can see from it, nearly the whole of the State of New Hampshire, with its numerous mountain peaks. Eastward is Mount Washington, in solemn repose,—its neighbouring peaks of immaculate whiteness—Mount Lafayette and its lines of white extending far down into the evergreen forests. Southward is Lake Winnipiseogee, with its numerous isles, glittering in the sunlight, like a gem of the purest water. Westward is the whole State of Vermont, and Ascutney, the most pointed of its mountains, is conspicuous. Moosilauke is so much higher than the immediate neighbouring peaks, that the whole country is spread out as a grand intrusive raised map before the beholder.

"No scene more grand and beautiful ever greeted the eye of man, than when, beyond the dark band of clouds just below the summits of the Franconia and White mountains, appeared those tints of rose and orange, lying along the horizon just above the snow capped summit of Mount Washington, and against a deep azure sky. From Moosilauke, you command the whole panorama of the White Mountain range, and you may see something of the effect witnessed among the Alps. As the day dies, the lost shadows pass with strange rapidity from peak to peak, vanishing from one height as they appear on the next."

The following are extracts from their Journal, viz:

"On the 1st of January, 1870, the sun rose clear. We were above the clouds, and a grander spectacle one does not often behold. The clouds seemed to roll and surge like the billows of the ocean. They were of *every dark and of every brilliant hue;*

here they were resplendent with golden light, and there of silvery brightness; here of rosy tints, there of sombre gray, here of snowy whiteness, there of murky darkness, here gorgeous with the play of colours, and there, the lurid light flashes deep down into the gulfs formed by the eddying mist. But above all these clouds, these flashes of light, this darkness, rise in stately grandeur, the summits of Mount Washington, sublime in its canopy of snow, and of Lafayette, with a few peaks of lesser altitude, glittering in the bright sunlight. As the sun rises higher, the picture fades away, the whole country is flooded with light.

"Did this grandeur, this magnificence, this brilliant display of lights, of shadows, and shades—of these clouds, so resplendent, so beautiful, portend a storm? In the evening the wind changed to the southeast, and increased in velocity.

"At daylight on the 2d of January, 1870, it was snowing. This soon changed to sleet, and then to rain, and at eight o'clock, A. M., the velocity of the wind was seventy miles per hour; at twelve o'clock, noon, there was a perfect tempest. Although the wind was so fearful, yet Mr. Clough was determined to know the exact rate at which it was blowing. By clinging to the rock, he succeeded in reaching a place where he could expose the anemometer, and not be blown away himself. He found the velocity of the wind to be ninety-seven and a half miles per hour, the greatest velocity, until that time, ever recorded. When he reached the house, he was thoroughly saturated with water, the wind having driven the rain through every garment, although they were of the heaviest material, as though they had been made of the lightest fabric. During the afternoon, the rain and gale continued with unabated violence. The rain was driven through every crack and crevice of the house and the floor of our room was flooded. So fierce was the draught of the stove, that the wind literally took away every spark of fire, leaving only the half charred wood in the stove, and it was with the greatest difficulty that we succeeded in re-kindling it. During the evening, the wind seemed to increase in fury, and although the window was somewhat protected, yet nearly every glass in it, that was exposed, was broken by the pressure of the gale. As the lights were broken, the fire was again extinguished, and even my hurricane lantern was blown out as quickly as if the flame had been unprotected. * * * * After nine o'clock, P. M., there were occasional lulls in the storm, and by midnight it had considerably abated.

"When it was clear, there was a strong temptation, notwithstanding the cold, to be out of doors to watch the clouds, at first of almost fiery redness, then changing to gray and neutral tints, until almost black, they seemed to gather around some distant peak, or as a dark band, they lay between the Franconia and White Mountains, leaving only the snow-clad summits above the dark border; or at sunset, when they lay in narrow bands, or rose tinted clusters around the summit of Mount Washington, while elsewhere they were those of leaden hue, such as are seen only in winter. Often when the sky is partially overcast, through the intervening spaces of the clouds, we see that intense blue sky, which is peculiar to high altitudes.

"On the 19th of February, 1870, there were two currents of air, the upper had its lowest stratum probably two thousand feet below the summit. In the morning the upper current was northwest, with a velocity of fifty miles per hour; about noon, the wind changed to the north and increased in velocity, and at five o'clock, P. M., it had a velocity of seventy miles per hour. At the foot of the mountain, nearly 5000 feet below there was scarcely a perceptible breeze, yet up, a thousand feet, there was a strong current from the *southwest*, and the clouds seem to move almost as rapidly as those from the north, higher up the mountain. On account of the velocity of the wind, and the upward pressure of the currents below, the effect was remarkable. The whole country, except the higher summits, was covered with clouds, and these were moving at the rate, probably, of more than sixty miles per hour, and everywhere they were broken into seething, undulating masses, for as they came near the mountains, in an instant, almost, they would be lifted more than a thousand feet, to be carried over the summits. As far as the eye could reach, embracing thousands of square miles, was this rolling tumultuous mass of clouds."

These gentlemen left the Moosilauke mountain on the last day of February, A. D., 1870. It was extremely cold, wind 60 to 70 miles per hour, thermometer ranging from 0 degrees to —17 degrees. The complete organization of the expedition to pass the winter of the years 1870 and 1871, on Mount Washington, was as follows, viz:

C. H. Hitchcock, State Geologist, J. H. Huntington, in charge of the Observatory upon the mountain. S. A. Nelson, Observer.

A. F. Clough and H. A. Kimball, Photographers.

Theodore Smith, Observer and Telegrapher for the United States Signal Service.

"Mount Washington, in the White Mountains in New Hampshire, is in latitude 44 degrees 16 minutes 25 seconds north and in longitude from Greenwich 71 degrees 16 minutes 26 seconds west, or 1 degree 0 minutes 43.99 seconds of longitude east from Hanover in New Hampshire.

"Its elevation above tide water is 6,293 feet, and in altitude it is the second highest mountain northward of the Gulf of Mexico, the highest mountain thereof being Clingman's Peak, in the State of North Carolina, which is 6,707 feet above tide water.

"The limit of the growth of trees on the north side of Mount Washington, is 4,150 feet above tide water.

"The climate of Mount Washington corresponds with that of the middle of Greenland, about seventy degrees of north latitude, or 26° further north than New Hampshire.

"It is an arctic island (so to speak) in the Temperate Zone, and on account of its great elevation it exhibits also the condition of the atmosphere, where the mercury does not rise above 24 inches in the barometer. For peculiar interest therefore, the Mount Washington Station is not exceeded by any point within the arctic circle."

Professor Edward Tuckerman, of Amherst, Massachusetts, in his admirable treatise upon "the Vegetation of the White Mountains," marks out four regions: first, *the lower forest*, in which are found the hard wood species of trees, the rock maple, the beech, the white and yellow birches; with these are often large white pines, firs, white spruces, the aspen, the witch hazel and the mountain ash.

"In the second region, *the upper forest* consists mostly of black spruce and fir, with occasional yellow and canoe birches, Frazer's balsam fir and a mountain ash; at 4,000 feet cf altitude these trees are dwarfed but are very strong, and when close together form a thicket almost impenetrable.

"Among the plants of the third or *sub-alpine* region are the mountain sandwort, the evergreen cowberry, the Labrador tea and the mountain bilberry. This seems not to be well characterized.

"The fourth and highest region is called *alpine*, and contains many plants peculiar to Labrador and Greenland. There are some fifty or sixty of these, and among them are as many more lowland species which have emigrated to the summit and manage to live there in favourable seasons, though often much dwarfed. The lichens are very conspicuous and beautiful, one of a sulphur yellow colour is quite noticeable, and is a good indication of the visitor's arrival in the Alpine District. Another is the reindeer moss, a very common article of food for the most useful animal to man in Lapland. The best localities of these arctic plants are in the great gulfs or ravines upon the east side of Mount Washington.

"As far as the upper limit of trees, boulders that have been transported by the glacial drift from more northern summits are common. They rapidly diminish in number and size upon that point, and have not been seen far above the fourth water-tank, or above an altitude of 5,800 feet.

"It is winter weather on Mount Washington in October. Most of the necessary preparations having been made on November 12th, 1870, Mr. Huntington promptly climbed Mount Washington and commenced to take and record the meteorological observations. The other members of the party were delayed by various reasons—but on the 30th of November, 1870, four gentlemen of the party, viz: Charles B. Cheney, of Oxford, A. F. Clough, of Warren, C. F. Bracy, of Warren, and Howard A. Kimball, of Concord, arrived at the summit, and on the 4th of December, 1870, Sergeant Theodore Smith, of the U. S. Signal Service, detailed as an observer, joined the party.

"November was making its exit in what might be termed a lovely winter day, and the prospect of so choice a time to make our ascent, toilsome at best at this season, and very hazardous except at special times in good weather, inspired us with enthusiasm more and more increased as we approached the final reach that stood in defiance of any aid that could be rendered by the panting steeds that now bore us forward.

"At Marshfield we are three miles from the summit, and at present all travel over this distance must depend solely upon human muscle and energy to achieve. At this point we decided to make the ascent at once, though there were serious misgivings on the part of some of us in view of the near approach of night, which at this season, half-past two o'clock,

P. M., leaves a small margin of the day, at best for such a task as stood before us. In ascending from this point we followed the railroad track. We were compelled to walk upon the ties for the snow was several feet deep, with a sharp upward grade in some places rising one foot in three, with the ties three feet apart and loaded with ice and snow and built on trestle work over gorges of some 25 or 30 feet in depth; the careless eager steps of unbaffled enthusiasm, are soon compelled to give place to great caution and the constant stress of nerve and muscle. * * * * The end of the first mile carrying us up to within one half mile of the limit of wood growth, found us in tolerable condition, when a halt for breath and observations discovered to us an approaching storm lying on the Green Mountains of Vermont. It would undoubtedly strike us but we still hoped we might press on and reach the summit first. The thought of being overtaken by a furious storm on the wintry, shelterless cliffs of Mount Washington, with the night about to enshroud us, was fearfully impressive, and prompted us to our best endeavours. With all the effort we could well muster, we had only advanced a half mile more, carrying us fairly above the wooded region to the foot of 'Jacob's Ladder,' when the storm struck us. There were suddenly wrapped around us dense clouds of frozen vapour, driven so furiously into our faces by the raging winds as to threaten suffocation. The cheering repose of the elements but a moment before, had now given place to what might well be felt as the power and hoarse rage of a thousand furies, and the shroud of darkness that was in a moment thrown over us was nearly equal to that of the moonless night. Compelled to redoubled efforts to keep our feet and make proper advance, we struggled with the tempest, though with such odds against us that we were repeatedly slipping and getting painful bruises. Mr. Kimball finding himself too much exhausted to continue this struggle on the track, we all halted in brief consultation—during which Mr. Clough suggested that our only hope consisted in pushing upward with all our might.

"Here we became separated, three of our party left the track, and Mr. Kimball willingly left behind his baggage in order to continue the ascent. By thus leaving the track, we escaped liability to falls and bruises, but found ourselves often getting buried to our waist in snow, and forced to exert our utmost strength to drag ourselves out and advance. We repeatedly called to Mr. Bracy, who had kept on the track as we supposed, but could get no answer. The roar of the tempest overcame our utmost vocal efforts, and the clouds of frozen vapour that lashed us so furiously as it hugged us in its chilling embrace, was so dense that no object could be seen at a distance of ten paces. Against such remorseless blasts no human being could keep integrity of muscle and remain erect. We could only go on together a little way and then throw ourselves down for a few moments to recover breath and strength. We had many times repeated this, when Mr. Kimball became so utterly exhausted as to make it impossible to take another step. He called to the others to leave and save themselves if possible. The noble and emphatic 'never,' uttered by the manly Clough, whose sturdy muscle was found ample to back his will, aroused him to another effort.

"The two stronger gentlemen, whose habits of life and superior physical powers gave hope of deliverance for themselves, were both immovable in the determination that our fate should be one, let that be what it must.

"The situation was one of most momentous peril, especially as to Mr. Kimball, whose exhaustion was so extreme that he was wholly indifferent to the fate that seem to impend, only begging that he might be left to that sleep, from whose embrace there was felt no power of resistance. Still there was a listless drag onward mostly in the interests of his companions, and in obedience to their potent wills. After this sort we struggled on a few rods at a time, falling together between each effort to rest and gain new strength. At each halt Messrs. Clough and Cheney used their best endeavours by pounding and rubbing Mr. Kimball's feet and limbs, and in various other ways endeavoured to promote circulation and prevent freezing. The last saving device was supplied by a cord, which we chanced to have, and the end of this was made a noose, which was placed in Mr. Kimball's hand, while the other end was passed over the shoulder of Mr. Clough, who tugged along in advance while Mr. Cheney helped at his side. Most of the last mile was accomplished in this manner.

"With the wind at 70 miles per hour and the thermometer down to 7°, as was found after arriving at the Observatory, we came at length to 'Lizzie Bourne's Monument,' only thirty rods from the Observatory. One of our party shouted an exultant hurrah at the glad sight of this rude pile, which was erected to commemorate the sad fate of one who was overtaken by the darkness and bewildering fogs and chills of a rude October night. 'Then,' in the words of the eloquent Starr King, 'was the time to feel the meaning of that pile of stones, which tells where Miss Bourne, overtaken by night and fog, and exhausted by cold, breathed out her life into the bleak cloud.'

"It took more than a half hour's time to make this last thirty rods. Even the stronger ones had become wearied by their unusual exertions, and had not this been the case, their progress would have been slow, for it was found absolutely impossible to force on the one who had become unable to regard his own peril more than a few feet at a time. He would then sink down into a deep sleep, while the others would employ the time in chafing his hands and feet, and after a few moments manage to arouse him and make another struggle onward.

"From Lizzie Bourne's Monument to the summit, Mr. Kimball was mostly insensible to passing events, and only awoke to clear consciousness, as from a dream, to find bimself in bed in a comfortable room in the Observatory building, safe from the dreadful tempest, and owing his life to the unyielding devotion of these brave men who scorned to save themselves at the expense of a comrade left to perish. Mr. Bracy, who had got separated from us during our earlier struggles, had got in about 7 o'clock, P. M., our own arrival being at 7½ o'clock, P. M. He had kept on the track.

"Thus at least three hours of this ascent were made amid the darkness of a moonless night in the howling tempest, the horrors of which will be more readily appreciated when it is remembered that a wind of 45 miles per hour blew down buildings and uprooted trees in New York City. Twenty-five miles per hour added make a most fearful hurricane. We were abundantly supplied with nourishment on our ascent, chiefly in the form of a strong decoction of tea, of which we occasionally partook. This is found to be by far the most potent and effective stimulant that can be used in such conditions of extreme exposure.

"Mr. Huntington, aroused by the arrival of Mr. Bracy, sallied out with a lantern in search of us, but found his best exertions of little avail, the storm being so fierce and thick, he could neither make himself seen nor heard beyond a few paces, and they were regarding us as probably lost, though they were preparing for another effort in our behalf, when we arrived.

"A sleepless night gave place at length to a day thick and stormy, and for several days the clouds gathered densely around us, and the storm continued to rage, during which we were recovering from 'the wear and tear' of our adventures, and recruiting for the work in store for us."

The railroad depot, in a part of which this party passed the winter of 1871, was a wooden unfinished building, sixty feet long by twenty-two feet wide and stands nearly north and south. It has eleven feet posts and the elevation of the ridge pole is twenty-five feet, the roof of the usual form in ordinary buildings. The apartment occupied by the party is situated in the southwest corner of this building. It is a room about twenty feet long, eleven feet wide and eight feet high. The large part of the depot forms a sort of vestibule to this room, and is wholly inclosed except at the easterly end of the northern face, where the outer door is situated.

An extract from Mr. Kimball's diary, reads: "December 5th, 1870. The day is beautiful, we are perfectly comfortable outside without overcoats, and *on the east side of the Observatory, the frost is thawing quite rapidly*. Thermometer 22° Fahrenheit."

Now why, with the thermometer at 22°, should the thawing of the frost be confined to *the east side of the Observatory*, when the sun was shining all around the building on the snow or frost without thawing it elsewhere away from the building? If the thawing was the result of the heat rays of the sun, so improperly termed, why was not the thawing general all over the summit of the mountain, instead of being confined to one locality?

The explanation, I think, is this, viz: the early morning rays of sunlight being nearly horizontal, impinged with a velocity of 186,000 miles per second perpendicularly on the vertical wall of the Observatory, partly covered with frost work; great friction was produced by the impact and positive electricity evolved; this electricity rushing to the conjunction or embrace of the negative electricity of the frost work, when in contact with it developed heat which thawed the frost work over the other parts of the summit of the mountain; these morning rays of sunlight either passed horizontally or fell upon them with such small angles of incidence, as to be wholly reflected into the upper atmosphere.

Mr. Kimball continues: "we have succeeded in making some

very good (photographic) views, but not as large a variety as we intend to have before we complete our winter's work. * * * We have also made three negatives of clouds, which were at least half a mile below us. They resemble the waves on the ocean, only the cloud waves are in some places twenty or thirty miles long. They pass over a range of mountains, and take a long sweep across the valleys and then rise over the mountains on the opposite; and as a general thing, after passing over and coming down on the other side, they break up in small clusters resembling, on a grand scale, the surf from breaking waves. We have made some photographs of this. * * * * All these clouds move rapidly from the southwest, probably at a velocity of forty miles an hour, while on this summit, it blows generally from the northwest. We have made a view which shows a small portion of a remarkable cloud effect or phenomenon. It was like a parallel belt on the distant horizon, whose circuit must have been more than a thousand miles. It resembled the tire of an immense cartwheel, (we occupying the place for the hub,) which was beyond and encircled all the lakes, mountains, &c. It was even beyond Mount Katahdin—at the south, its upper edge was parallel with the point farthest north. At noon it appears to be approaching us as a centre, and as it nears us, it breaks up in magnificent great thunderheads, minus the thunder,—all this time our view is becoming more limited. * * * All this time it was snowing below, but we knew nothing of it until night. Our view of the surrounding mountains lasts only a short time longer, for we see to the west thick heavy clouds, marching upon us, and by 4 o'clock, we become densely shrouded—we cannot see Tip Top House from the Observatory not many feet distant.

"December 12th, 1870. This morning the wind was south, but changed to the northwest in the afternoon; at ten, A. M., there was a bow in the clouds, and at noon there were in addition three supernumerary bows which remained for an hour and a half, and some of the time they were remarkably distinct. Late in the afternoon the sky was intensely blue."

From their journal we make the following extracts, viz:

"December 21st, 1870. Messrs. Kimball and Thompson (a visitor,) took an observation from the roof of the Tip-Top House; wind 60 miles per hour. They were out but five minutes, yet their coats, caps and hair were covered with frost and Mr. Thompson had slightly frozen a finger. Later, the wind had fallen to 30 miles per hour, and now, eleven o'clock, P. M., it is moderate for Mount Washington.

"1870, December 23d. A cold morning, thermometer zero, but we don't feel the cold as sensibly as in the lower regions.

"December 24th. Yesterday afternoon and late at night a 'snow bank' lay along the south; this forenoon, snow was falling with a temperature of —13°, at times during the day the wind was as high as 70 miles an hour, consequently, we were confined to the house. It is cold to-night, (now nine o'clock, P. M.,) the thermometer —15°, and only 42° in the room, although we have two fires.

"December 25th. There were no clouds above or around the summit. Below, and but a little lower than this peak, the clouds were dense and covered an extensive tract of country. Through the less dense portion of the lighter clouds the sun's rays gave a peculiar rose tint, extremely beautiful in effect. * * * * About ten o'clock, A. M.,. Mr. K. and myself went out for an observation. We had the pleasure of witnessing the formation of several coronæ, sometimes single, but oftener three; even on one occasion *four* distinct circles appearing and disappearing so rapidly that it was impossible to more than catch a glimpse of form and colour. It was a phenomenon of rare beauty.

"December 29th, 1870. The wind has been increasing all day. At 7 o'clock, A. M., observations: wind, 46 miles per hour; at 2 o'clock, P. M., 57 miles; at 4 o'clock, P. M., 72 miles; at 7 o'clock, P. M., 46 miles; and at 9 o'clock, P. M., nearly calm; a great change in 14 hours, especially in the last two hours. Barometer has fallen rapidly all day.

"December 30th, 1870. The morning is calm, clear and beautiful. It is what we have waited a month for. We commenced work making negatives at sunrise. In the morning we made a few 8 by 10 negatives, but as we were making the last of them the wind freshened up, and we could not make as many as we wished. * * * Before I close to-day's memoranda I must speak of the splendid view we had after the wind, by blowing so fiercely, obliged us to quit work. We could see distinctly hundreds of mountains, lakes, ponds, &c. Off to the northeast in the distance—one hundred and fifty miles distant —we see Mount Katahdin, the highest mountain in Maine, and a little to the north we see mountains which apparently are much farther away than Mount Katahdin, and must be in the upper part of Maine, near Canada. We never before saw the ocean nearly as plain as to-day; we could see a great distance 'to sea.' Off to the southwest we could see Kearsarge mountain and Monadnock, and over the Green mountains, the Adirondacks and Lake Champlain, in northern New York, were distinctly visible. About 2 o'clock, P. M., I noticed a long hazy line over the ocean; soon it grew larger and then I could see it was nearing us, and in an hour it was within 40 miles, and we could see it as a vast sea of cumulus clouds. The wind was increasing, and had changed from the east to the south, and it carried the approaching clouds and storm to the north of us. We were thankful to see it go by without striking us, for it is grand to behold but not desirable for a covering. To-night we have some of the effects of it in the wind, which, as I write, is blowing a most violent hurricane, making the Observatory creak. A few hours ago the wind was scarcely noticeable; now its velocity is over eighty miles an hour, and for a wonder it comes from the south, instead of northwest as usual, and as a natural consequence it tears off all the loose ice and frost from the Observatory. It seems as if we were at sea in a severe gale, and broken ice and timbers were beating against our ship, and at times our building shakes like a vessel in a storm. Contrary to what ordinary experience would seem to teach, the north side of the building is less exposed to the fury of the element than any other." This is owing to its having but one electricity.

Now, why does not the north wind, or the northwest wind, produce similar effects? The sun shines upon both winds alike, and if it sends down heat to this planet, the northwest wind should be as warm as the south wind, and should tear off the frost-work from buildings and rocks just as the south wind does. But no such effects are observed during the prevalence of these northern winds; on the contrary, it is only while these northern winds are blowing in winter that this frost-work is formed.

The explanation I conceive to be this: the southern winds coming from a warm atmosphere are positively electrified, and when they reach the frost work on the buildings or rocks oppositely electrified, their impact produces friction, which evolving more positive electricity, develops heat that detaches the frost work from its adhesions, breaks it into pieces, and finally melts into water—while other frost work protected from the south wind remains firm and unaffected, the temperature of the atmosphere being below the freezing point of water. "A telegraphic wire connected the Observatory with Marshfield, a distance of three miles, where it is joined with the Western Union Company's line, at Littleton, twenty-three miles farther. The wire has frequently been charged with atmospheric electricity, especially in the afternoon of the 7th of January, 1871, when, on account of the high tension of these currents, it became utterly unmanageable. When the key was opened, the flow of the current still continued, exhibiting bright sparks, leaping from one platinum point to the other. After dark, no auroral display could be seen. There is also a wire connecting the summit with the Glen House, which is detached from the poles and laid upon the ground during the winter, to protect it from the violent winds prevailing at this season. We had it attached to an instrument, and, although no battery was used, we discovered that it was sometimes charged with electric currents, which deflected the needle considerably. The Glen wire was broken about a mile and a half from the summit, and the one down the railway had parted at about the same distance, thus making the phenomenon quite remarkable.

"1871, January 10th. After ten, A. M., the summit was free from clouds, but below masses of clouds were driven along the valleys and over the lower summits. The clouds about and over gave grand effects of light and shade along the mountain range—they were particularly fine on Adams and Jefferson and near the Glen. The snow is nearly all off the houses and the rocks—a great change in three days' time. I cannot let this day pass without a mention of the high temperature; at one o'clock, P. M. it was 37°. Like April it seemed, but who knows what it will be to-morrow?

"January 14th. Last night we saw a fine aurora, broken arches with streamers, never before was one apparently so near; it certainly did look as though it was within reach.

"January 16th. Still raining; at eleven o'clock this forenoon, Mr. S. started out on a voyage of discovery, but it rained so hard and the walking was so difficult that he soon came back. * * * Mr. H. went down to the spring to-day and brought up a pail of water. A week ago this was an arctic region, now it is more like April in the valleys of New Hampshire.

"January 17th. The wind was high during the night, say eighty miles per hour; at 7 o'clock, A. M., to-day, only 75 miles per hour, strong enough however to compel Mr. H. to sit while he measured the force of the wind that he might not be blown over into Tuckerman's ravine. * * * * Has blown stiffly all day, yet we have taken the air several times; pleasant walks in the face of a fifty mile breeze. Perfectly clear at sunset. Had one of the best views of the shadow of Mount Washington on the sky yet obtained. The mountains far and near look dull and gray now since the rains.

"1871, January 19. Mr. H. called us out, before sunrise, to see the beauty of the morning; in truth it was wicked to miss such a glorious view as we had. Perfectly clear, and nearly calm. Never before have I seen the shadow of the mountain so grand on the western sky, never so charming the purple tints at break of day. Never so impressive have been the shaded outlines, the lights and shadows on the mountains and in the valleys, as on this memorable morning. Sunset was but the complement of the morning, and the evening is beautiful as ever night can be, the stars shine with a light as soft as June, all, all is beautiful.

"January 22, 1871. Having a gale to-day, and not only a high wind but a temperature below any thing I have ever experienced before; now, at nine, P. M., —34 degrees inside the door; at two, P. M., wind 72 miles per hour. Professor H. measured the velocity, he had to sit with a line around him, myself at the other end indoors as an anchor; even then it was impossible for him to keep his position. Temperature —31 degrees. I put up a pendulum, this morning, in our room, it is four feet long, and the rod passes through a sheet of cardboard, on which are marked the points of the compass. The oscillations, when the wind blew in gusts, were in every direction, changing suddenly, and sometimes had a rotary motion. When the wind was steady, the oscillations were northwest and southeast. With two fires the room is cold to night.

"January 23, 1871. The wind raged all night. The house rocked fearfully, towards morning the wind ceased, and all day it has been nearly calm. The temperature outside —43 degrees. Professor H. and myself sat up all night to keep the fires going. The pendulum gave oscillations of an inch and a half at times during the night. Temperature to-night at ten o'clock—40 degrees; a changeable climate this.

"January 31, 1871. The most glorious sunrise this winter. To the east was a sea of clouds broken and much lower than usual. The protruding peaks resembled islands, more than ever before; over northern New Hampshire and Maine, and along the coast, the clouds were very dense, but their upper surface, as the sun shone across them, was of dazzling brightness, while singular forms of cirrus clouds overcast the sky. Low in the west it was intensely black, and detached masses of clouds floated along the northern horizon. For an hour after sunrise all these cloud forms were constantly changing in colour—purple and crimson, leaden hues and rose tints, almost black and dazzling white.

"February 1, 1871. Clouds on the summit till noon, when it suddenly cleared up. Early in the forenoon, the wind was fully 50 miles per hour, at noon it was nearly calm, and till nine, P. M., not above 20 miles per hour. At nine, P. M., the thermometer indicated —16 degrees.

"From 3.30, P. M., to sunset, there were the finest cloud displays possible. Eastward, heavy masses of clouds, in color from gray to an intense black. Westward, detached cirro-stratus, presenting every shade and colour; along the northern horizon a clear light rested; the west was burning bright in crimson, purple, and gold, while far south, fading out toward the east into gray, the colour was a delicate rose tint. Below, to the west, far as we could see, the whole country was covered with cloud. The icy peaks glow and glisten in the bright sunlight. The transitions of shades and tints, the colours burning into the radiant sunset, surpassing any thing we have seen yet for a sunset scene, mark this as a day never to be forgotten—as I write, it seems like a dream.

"1871, February 2d. All day the wind has been light, and it was nearly calm this evening till half an hour since, when, without any warning, (except the falling in the barometer,) the gale began, not with a rising wind, but with a single blast that shook the house to its foundations. * * * Now, at 11 o'clock, P. M., the wind has risen to the dignity of a gale. The temperature —20° out of doors.

"Friday, February 3d. Well, it did blow last night, making some of the time such a racket out-doors and in-doors too, for that matter, that sleep was out of the question. The wind must have been as high as 90 miles per hour during several of the heaviest gusts. For a change to-day, we get the most severe snow storm of the winter so far. The wind is northwest, the point from which our storms and hurricanes come. At no time has the temperature been higher than 5°; it was —25° this morning at 7 o'clock.

"Saturday, February 4th, 9 o'clock, P. M. The wind rising toward morning has held its own all day, at no time being below 75 miles, and since 8.30, acts as if it was ambitious to attain the 90 miles per hour standard. At 7 o'clock, A. M., temperature —33°; from 5 o'clock, P. M., to this last observation it has gradually worked down to —40°. We have not suffered from the cold simply because we have not exposed ourselves to it. In the room at no time has the temperature been lower than 34°, and most of the time we have managed to keep it up to about 60°. To do this, we have the stoves at a red heat; the thermometer hangs precisely five feet from the stove; ten feet from the stove at the floor, to-day, the temperature was only 12°, at the same time was 65° in other parts of the room. Midnight—really there is quite a breeze just now. Some of the gusts, from what we know of the measured force, must be fully up to 100 miles per hour. In fact it is a first-class hurricane. The wind is northwest, and as the house is broadside to it, the full force is felt; at times it seems as if every thing was going to wreck. We go to the door and look out; it is the most we can do; to step beyond, with nothing for a hold fast, one would take passage on the wings of the wind in the direction of Tuckerman's ravine. However unwillingly one might go, such would be the result if he should venture outside, so irresistible is the force of the wind. What varied sound the wind has as it changes, now howling, screeching, roaring as though the building was surrounded by demoniac spirits, bent upon our destruction. We shout across the room to be heard. Now it suddenly lulls, and moaning and sighing it dies away; then quickly gathering strength, it blows as if it would hurl the house from the summit. The timbers creak and groan and the windows rattle; the walls bend inward; and as the wind lets go its hold, rebound with a jerk that starts the joints again. The noise is like rifle firing in fifty different directions, at the same moment in the room—a moment ago close by me as I sat here, leaning against the wall, now in the outer room or up aloft and outside as well. Then there is the trembling and groaning of the whole building, which is constant. Everything movable is on the move, books drops from the shelves, we pick them up, replace them only to do it again and again. The temperature is now —10°.

"Sunday, February 5th. From one to two o'clock, A. M., the wind was higher than during the early part of the night. Some of the gusts must have been above 100, possibly 110 miles per hour. The tempest roared and thundered. It had precisely the sound of the ocean waves breaking on a rocky shore, and the building had the motion of a ship scudding before a gale. At 3 o'clock, A M., the temperature had fallen to —59°, and the barometer stood at 22.810, attached thermometer 62°. Barometer was lowest yesterday at 8 A. M., when it was 22.508, and attached thermometer 32°. Now, 7 A. M., the thermometer indicates —25°, and the wind has fallen to 70 miles per hour. By accident, the spirit thermometer has not yet been received. But this has been the only day when the mercurial instrument has not been perfectly reliable. The valleys are full of stratus clouds; charged with frost as they are, occasionally sweeping over the summit, they completely cover one in a moment, hair, beard and clothing; when the face is exposed it feels like the touch of hot iron. To breathe this frosty air is very unpleasant. A full inhalation induces a severe coughing fit.

"Monday, February 6th, 9 A. M. Talked over the events of the past night at the breakfast table. * * * Of all the nights since this party came here, the last exceeds every one. 9 P. M.; it has been a rough day, down in the world people would say a severe one, so should we but for the recollection of last night; our coal bin is under two feet of snow, and anywhere in the room, that snow is six inches deep. The highest temperature is to-day 12°, and the lowest now, at 9 o'clock, P. M., is 2°, a very acceptable change—wind 50 miles in the forenoon, now 20 miles per hour, is good as a calm. It is clear, and the moonlight is that of the mountain, seen only at this or higher elevations.

"Tuesday, February 7. A glorious sunrise; a quite warm day, and at sunset almost equal to that of the 1st; temperature at 2 o'clock, P. M., 62° in the sun; change of temperature since Sunday of 121°."

These sudden and great variations of temperature in the same latitude elevation above the sea, and identical locality, in short spaces of time, are strong evidences that the temperature of our atmosphere is exclusively to be attributed to electrical causes within it, and not to any supposed rays of heat emanating from the sun.

"Tuesday, February 7th. I have given some time this afternoon to the study of cloud formations. Days like this are so rare that we improve every opportunity for investigation. Gales, storms, hurricanes, all clear off with a north wind—a wind gentle and soft as the south wind of the lower regions. How can this be explained? It is S. S. W. to-night and 2 miles per hour; a marked contrast to Sunday morning.'

Let us attempt an explanation of this phenomenon: When masses of clouds, freighted with moisture, and at different elevations, approach each other, attracted by their opposite electricities, heat is evolved by their conjunction. The watery vapour constituting the clouds undergoes a radical change; the atmospheric air, which holds the water in suspension, absorbing the heat that is evolved by the conjunction of the opposite electricities of the clouds in commixture, is so greatly expanded and rarefied that its molecules can no longer sustain the particles of water with which they had been associated; this attenuated air, thus heated, leaves the watery particles, and being positively electrified, is attracted by the opposite electricity of the higher atmosphere and ascends instantly into it, while the water being negatively electrified is repelled from the air above, and begins to fall in sheets, which soon separate into drops, repelling each other, and carrying to the earth the electricity in a latent form with which they were associated. When the clouds have thus discharged all their water as hail, snow or rain, to the earth, the atmosphere in which the floated becomes very dry and electrical. The north wind, warmed by the heated air which has escaped from the clouds when they met, is attracted to the spaces before occupied by the clouds in the direction of the ocean and becomes the gentle, balmy air described by these observers, and as dry air has an electricity always opposed to that of moist air, the north wind at Mount Washington always is attracted to the Atlantic ocean to the south of the mountain, and storms thus terminate in that locality with a north wind.

"Wednesday, February 8th. Ten o'clock, P. M. There is evidently a snow storm along the coast, the northern edge, within fifty miles of us. This forenoon we could see the storm as it moved eastward. It was cloudy and clear by turns on the summits, that is, the lower current of cloud rested at times over us. The valleys east were full, and the upper stratum overcast the entire country as far as could be seen. Wind S. S. W., from 20 to 50 miles per hour. Temperature from

14°, at 7 o clock A. M., to 20° at 2 P. M. Interesting to watch the progress of the storm and to see the lower current of cloud driven by an easterly wind, running under the higher stratum which of course is *toward* the northeast.

Let us here stop to admire the infinite wisdom of the Creator, who, using the attractive forces of his electricities to gather and collect the watery vapours of the atmosphere into clouds, disperses them by the repellent forces of these same electricities and scatters in this way their manifold watery blessings over greatly increased areas of the surface of the earth.

"Thursday, February 16th. A storm of snow and rain. It rains here, with the thermometer at 22°, as it did to-day, and snows with it at 30°, as might be expected. Why it should rain at 22° is hard to explain. Wind steady; southwest through the day; but, at 8.20 P. M., changed suddenly to northwest in gusts, 60 to 80 miles per hour. Forgot to mention last night, that at 6.30 P. M. I read from the 'Atlantic' in the open air. Our days are about 46 minutes longer than they are at the sea level."

The warm southwest wind explains the rain at 22°, which was probably the temperature outside of the column of warm air brought up by the southwest wind.

"Sunday, February 19th. A bright, sunny day, clear and calm, yet the temperature was at no time higher than 8°." Where was the sun's heat?

"Tuesday, February 21st. When S. left this morning the thermometer read —4°, and wind 20 miles per hour; at the Gulf Tank it was so warm he had to lay aside overcoat and gloves; no wind there; the snow was melting and the water running down the centre rail; quite a contrast to the summit, only one mile distant—meteorologically speaking, he was 300 miles south of his mountain home, though in sight of it. We took a walk. Fine weather for a change. Beautiful cloud views this afternoon. Light fleecy clouds floating over Mount Monroe. Dissolved before reaching Tuckerman's ravine. They passed between us and the sun, showing the prismatic colors; then as they rolled eastward, gradually faded out and changed to a cold gray. The transitions of light and shade were inexpressibly beautiful, enough to give sensations of pleasure to the dullest observer, and drive an artist crazy with delight.

The buildings are cased in ice and frost work of most elegant forms, resembling rocks, flowers, leaves, shells and the wings of birds.

"February 24th. From 9 o'clock A. M. to 3 P. M. the temperature varied but a degree or two from 37°; the barometer steady.

"February 27th. This time we are favoured with a rain storm, pouring when it was calm, and in driving sheets after the wind rose to 84 miles per hour. At 9 A. M. it changed to snow, and then, by turns, rain for a moment, then quickly changing to snow, and suddenly rain again; but the snow obtained the mastery.

"February 28. It cleared off early in the morning. Wind from 50 to 70 miles per hour. The mean temperature, zero.

"March 3. A storm seemed to be brewing last night at a late hour, and early it came, a heavy rain storm. Towards noon the wind rose, and at one P. M., it blew 96 miles per hour. How the wind roared in the flue! How the house shook! Had to shout across the room to be heard. It was grand, however. From 4 o'clock P. M. the wind abated.

"March 23. At 9 P. M., snow squalls to the northeast, and the clouds gradually settling in the valleys. * * * * By 2 P. M. the mountain was in the clouds. They were at a higher elevation than has generally been the case—cirro stratus; color gray; uniform in density nearly over the entire field of view. * * * * Evidently the lower current was from the east, while the wind on the summit was west northwest. * * * * The clouds passed over Mount Adams, and later over the dividing ridge, between Mounts Washington and Clay. They seemed to curve, as they passed over these mountain tops, as though the upper currents of air conformed to their irregularities of surface." [The mountains and the clouds having the same electricities, which repelled each other.—*The Author.*]

"When there are two strata of clouds, they unite before the snow or rain falls, as a rule, though to-day the snow fell an hour previous to the clouds settling on the mountains.

"April 1. To-day, 64 degrees in the sun, at 11 A. M. Afterwards cooler—15 degrees at 9 P. M. * * * * A northeast wind to-night, seldom from that quarter.

"April 3. * * * * Such is the atmosphere here, that although the thermometer, in the shade, marked 27 degrees, I wore neither hat nor coat, and yet was warm enough.

"April 4. All the forenoon, till one P. M., the summit was in a dense cloud. Suddenly it lifted, and then we had the most gorgeous display of cloud-scenes we have yet witnessed. Eastward, masses of cumuli rested over the valleys and the mountains. Why not call them mountains of clouds? Certainly. They rose far above our level, six thousand or perhaps eight thousand feet higher than this peak. They conformed to the heights over which they lay, and seemed to envelop other mountains, nearly as lofty as their upper limits. The illusion was perfect, and Mount Washington in comparison, was a diminutive spur, or outlying peak of this great mountain range. * * * * The sun rises high, but we know nothing of Spring. Truly it is more like Winter than some of the time in March. Then there was no snow. Now, everywhere there are snow and ice.

"April 6. A clear sunrise—cold—thermometer only 3 degrees, the wind 20 miles per hour, and the morning view, that of December. Though clear, the sun gave little heat—a pale white light; the sky a light blue, and so clear, that it seemed almost as though we could see beyond its bounds, or through it into the regions of space.

"April 15. The rule holds good; no two days alike on Mount Washington. Ten hours we had splendid cloud-effects in every direction; cumuli north, in every form beautiful and fantastic, and colors as though some radiant angel had thrown aside his robe of light.

"April 28. To show the changes in temperature here, in a few feet of altitude, I note my trips down, to-day, and up as well. Left the house at 4.30 P. M.; wind 30 miles an hour; at the Lizzie Bourne monument, 40 miles; at the Gulf House ruins and below, 60 miles, thus reversing the order of things in regard to wind. Thermometer on the summit 28°; frost-work forming some distance below the monument. At the Gulf Tank, when the sun came out, as it did several times, the ice on my cap would thaw completely; then, while the cloud was passing, icicles two inches in length would form on the visor. It was difficult to work or even stand against the wind below the Gulf House ruins. Returning, the wind was not so violent.

"May 1. May Day, and still it is winter; every aspect is that of mid-winter. The spring near the Observatory remains frozen solid, and so we daily melt ice for use, and yet down the mountain a half mile there is seldom a day when the streams are not running.

"May 4. Another tough snow-storm; * * * wind got up to 48 miles per hour and temperature down to 21°.

"May 5. The storm—snowing in such a wintry way last night—turned to rain toward morning, and has been raining all day. * * * The wind was west here—not higher than five miles per hour—yet in the valleys it must have been much stronger, judging by the velocity of the clouds; besides, we could hear distinctly its almost roar.

"Monday, May 6. This morning clear, calm and warm. The thermometer, at 8 o'clock A. M., indicated 85° in the sun; warmest morning this spring.

"May 7. The barometer fell 50-100ths from last night at 9 o'clock to this morning at 7 o'clock. Wind rising at 3 o'clock A. M.; reached its highest velocity, 67 miles per hour, at 2 o'clock P. M.—highest recorded for some time, quite strongly reminding us of the winter months. Snowing all day. * * * At 5 o'clock P. M. the cloud passed off and we could see that not the mountains alone, but the lower country as well was 'snow-bound.' At 9.40 P. M. snowing again. Temperature, 2 o'clock P. M., 21°—highest for the day—and 19° at 9 o'clock P. M.

"May 8. We did have a rough night; called tne wind 80 miles per hour at midnight. Temperature at 7 A. M., 15°.

"May 9. Mountain peaks white as winter, but the valleys are bare. The frost work has seldom been more beautiful. Measured some feathers to-day, on a tall pole, at the Tip-Top House; found them 36 inches in length, and on a rock south of the house 49 inches in length and 15 inches broad.

"May 11th. A wintry sky and winter scenery this morning. The sky a pale blue and the sunshine that of December. * * * * Temperature 20° at 7 o'clock A. M.

"May 14th. The wind was high as 80 miles per hour, if not higher, during the night. All day, as usual, it has been cloudy and frost work forming. Temperature at 7 A. M. was 11°, and highest for the dayat 9 P. M., 21°; at no time the wind lower than 46 miles per hour. Mr. H. left at 9 A. M. in the face of a 48-mile gale and the temperature only 14°. I am anxious for his safety, and shall be till S. returns.

"The winter's work is done. Storms of unparalleled severity, when, for days in succession, the summit was enveloped in clouds, and the hurricanes lasted longer and were more violent than any yet recorded in the United States, together with very low temperatures, have been a part of our experiences. Just such an experience has seldom before been the lot of human beings. * * * And ours has been the good fortune to witness some of the most magnificent winter scenery upon which mortal eyes ever rested, scenery of transcendent grandeur and views surpassingly beautiful.

"There were days when the shifting views of each hour furnished new wonders and new beauties, in the play of sunlight and changing cloud-forms, every hour a picture in itself and perfect in details. Sunsets, too, when an ocean of cloud surrounded this island-like summit, the only one of all the many high peaks visible above the cloud billows, all else of earth hidden from sight; there were times when this aerial sea was burnished silver, smooth and calm, and times when its tossing waves were tipped with crimson and golden fire. * * * * Gone are the long days and longer nights, when the stoves failed to comfortably warm the little room, though we kept them at a red heat, and when the thermometer indicated 65° near the stove and 4° at the floor ten feet distant."

We have presented these extracts from the published observations of the gentlemen who passed the winter of the years 1870–1871 on Mount Washington, to show the sudden and great variations of temperature that occurred on the mountain by day as well as by night, and that these variations could not have resulted from solar radiations of heat, as sometimes when the atmosphere was the clearest and freest from vapour, and when the sun was shining with the greatest brilliancy, the temperature on the mountain was lower than when these conditions of the sun and atmosphere did not exist, and further, when the sun had passed the vernal equinox, and was approaching the summer solstice, the temperature on the mountain, and the condition of its atmosphere, continued still to be wintry, unaffected by the change in the position of the sun, relatively to the angles of incidence of its rays.

When we consider the altitude of Mount Washington,

which is only 6,293 feet above the sea level, or not much more than one mile, we find that its projection above the periphery of the earth would be about 1-8000 part of the earth's diameter, a protuberance so slight as to be wholly inappreciable at the sun's distance of 92,000,000 of miles from it. What proportion of solar radiation of heat (if there is such a thing,) could fall upon so microscopical a spot as Mount Washington, cannot therefore readily be imagined. But when we contemplate the electrical forces of our planet developed by sunlight, the radiation of interior terrestrial heat into the atmosphere—the movements of oppositely electrified currents of air, and the commingling of tumultuous masses of cumuli clouds, all evolving heat and changing with great suddenness the temperature of various localities, we begin to comprehend the plan of the Creator in furnishing each planet with its own sources of heat, instead of attempting to supply them with heat through almost interminable spaces, from so distant an orb as the sun. To an observer outside of our atmosphere, looking down upon our planet, he would see sometimes masses of dense clouds, which, intercepting the sunlight would cast dark shadows of various forms and sizes proportional to the clouds which would form them on the surface of the earth. The darkness of the shadow would be in proportion to the depth and density of the clouds floating between the sunlight and the earth. These shadows would flit across our earth as rapidly as the clouds which had produced them, in great storms or hurricanes of perhaps 100 or more miles per hour. Now may not the sun spots which have so much exercised our astronomers be produced in a similar way? Clouds or vapours of various luminosity being interposed between the most luminous part of the sun's envelope and the gray atmosphere of the sun, would cast upon the latter shadows so dark and so flitting as to resemble the shadows of clouds on our own planet, and the dispersion of the clouds so making the shadows would account for the rapid disappearance of the sun spots. The forms of the sun spots would vary with the sinuosities and unevenness of the surface of the gray envelope of the sun upon which these shadows fell, and the continual interference of intense light derived from other luminaries of the stellar world, with the fainter light received from our planetary system, would greatly increase the darkness of the shadows so produced.

Let us now consider the case of a total eclipse of the sun by the moon. In the reports of observers, the following appearances have been described: Solar prom nen es during eclipses, red protuberances, red clouds, red flames, &c.

One observer says: "They form around the solar globe a denticulated and continuous series of projections of very curious appearance." Another observer says: "The prominences were seen very distinctly, their colour was that of red coral, slightly tinted with violet. They all appeared to be adherent by their bases, and none of them floated detached at a certain distance from the moon as was observed in the years 1851 and 1861.

"The following facts may be considered tolerably certain:

"1. The prominences (or protuberances) belong decidedly to the sun.

"2. The prominences are of a gaseous nature, that is, they are composed of an incandescent gas, principally hydrogen gas, but they contain doubtless other substances, perhaps substances that are unknown on the surface of our earth, at least such would appear to be proved by the existence of a brilliant line in the spectrum, near to the yellow line of sodium, but not coinciding with the latter, and, moreover, most curious to relate, it does not coincide with any dark ray of the solar spectrum.

"3. The matter which forms the prominences is of very great extent, whether it spreads over the entire photosphere or not; it forms a continuous layer, the thickness of which is estimated by Mr. Loeyer, at some 5,000 miles on an average, and the prominences appear to be only portions of this layer projected to a certain distance from it, sometimes detached from it and floating above it. One of the great prominences represented upwards of 100,000 miles in vertical height above the photosphere.

"4. These stupendous accumulations of incandescent gas undergo, in very short intervals of time, very great changes in their form and size, which indicate that the layers of gaseous matter of which they form part are in a state of constant agitation, the cause of which is unknown, perhaps it is the same that gives rise to the spots and faculæ.

"It is extremely probable that the the entire globe of the sun has a very high temperature throughout its mass—a temperature which surpasses the melting (or boiling) points of most of the elementary substances of which spectral analysis has revealed the existence in its atmosphere. At the same time, it is evident that the various concentric layers of which the solar globe may be supposed to be formed, exert one upon the other considerable pressure, since we find that at the surface itself, the intensity of gravitation is twenty-eight times as great as it is upon the earth's surface. This pressure may hinder fusion to a certain extent, but not incandescence. But we believe that the hypothesis of a liquid incandescence or even a gaseous nucleus is more probable."

All such hypotheses are put at rest by the recognition of the sun as a great magnet, since magnetism is destroyed by heat.

"The prominences on the right, (western edge) appear like a mass of snow-capped mountains, the bases of which rest on the limb of the moon, and are lighted up by the rays of a setting sun." (From M. Jansen's observations on the eclipse of the sun from Aden to Malacca, August 18, 1868.)

"In 1858, M. Liais found that the light of the sun's corona, is really polarized, and at once concluded that the sun has an atmosphere extending far beyond the photosphere.

"During the short phase of total darkness, a luminous corona makes its appearance, being generally of a silver whiteness, but is sometimes coloured and surrounds completely the dark limb. Its apparent breadth is from one-fifth to one-twelfth of the diameter of the moon, and from it, light decreases gradually."

We have here in the aspect of the clouds in sunshine, from the summit of Mount Washington as they gather from the sea or from the land, advancing, stationary, or retiring, the most vivid descriptions of the varying brilliant tints and gorgeous groupings of colours, as the changing angles of incidence and reflection met their sight, that it is possible to conceive. We, who are familiar with the magnificent autumnal sunsets of many parts of our country, may begin to imagine the exquisite beauty of the scenes which these gentlemen have witnessed. But the particular object we have in view in calling your attention to it, is to trace the analogy of these displays of colour, light and shade, with those described by astronomers in investigating the physical condition of the sun. We have the same tints, brilliant colours, neutral colours, shades and shadows, in our planet as are described to be seen in the sun—similar disturbances in the vapour of both orbs.

Is it too much to imagine, therefore, that if an observer could be placed within telescopic range beyond our atmosphere, he might see in our atmosphere an exact imitation, upon a reduced scale however, of whatever has been exhibited by the sun, as the disc of our planet would then display a reflection of the illumination of the whole stellar world? And what more does the sun do? He receives the light of the whole stellar and planetary world, and reflects it again through space, thus presenting to one orb, or set of orbs, the light he has received from others, until throughout the great expanse, light is diffused everywhere to shine in the firmament of heaven, and give light upon the earth.

We have had exhibited in this city, (Philadelphia,) a few weeks since, by a distinguished artist, an oil painting of "Pike's Peak," one of the grandest mountains of the Rocky Mountain range. Its height is 14,216 feet above the sea level, and on its very summit is a signal station and observatory of the United States, erected in the year 1873. Its summit is covered with snow to a descent of perhaps a thousand feet. The painting, which represents a sunset scene, portrays the snow-covered summit, illuminated all over by a brilliant red tint, resembling red coral, and creating at first sight the impression of a mountain on fire. The resemblance to the red protuberances around the sun, during eclipses, as depicted in photographs taken by the observers, is most striking. This brilliant red coral colour pervades the whole surface of the summit of the mountain that is covered with snow, and which is seen through the red colour. Here we have an exact resemblance of one of the appearances of the sun, as displayed during an eclipse, and yet there is no incandescent gas covering "Pike's Peak" to produce this colour. On the contrary, the atmosphere around and above the mountain is wintry, with a temperature below freezing point "*Ex pede Herculem!*" May we not infer from this illustration that there is no incandescent gas about the sun, and that the varied tints and colours, however brilliant, and however resembling what we suppose to be incandescent metallic vapours, are really only manifestations of light in its protean displays, as fitful and evanescent as we see it in our autumnal sunsets.

Now let us for a moment imagine that by the interposition of the moon between the sun and the earth, each suffers an eclipse from the other. Let us suppose that the snow-clad mountains of our planet are bathed in sunlight, and that the brilliant colours derived from that source, changing with the angles of incidence and reflection, with which they encompass these snow-clad peaks, become displayed beyond the periphery of the moon, which has concealed a large part of the body of the earth. Now, if an observer could be placed between the moon and the sun, at the period of such an eclipse of the earth, would he not witness displays of light and colour, greatly resembling, if not identical, with those which would be seen by another observer placed between the moon and the earth, as he regarded the appearances about the sun? What then would become of the terrific heat of the sun and its incandescent gases?

"In the *hypothesis of undulations*, instead of supposing the transport of a material agent to great distances, it is held that the vibrations of luminous bodies are communicated to the atoms of an all-pervading *ethereal fluid*. These vibrations, propagated through this fluid, reach the organ of vision, which in time transmits them to the optic nerve. In this hpothesis, the nature and transmission of light would be analagous to the nature and transmission of sound, light being produced by atomic, and sound by molecular vibrations." This idea confines the action of light to animal vision.

In these cases there is no analogy, for sound has a very limited range of action, with comparatively small velocity, and is only of value to living beings. While light has scarcely a limit as to distance in penetration, and a velocity inconceivably great, and is indispensable to planetary existence.

Two persons hold a table-cloth, twenty-five feet long, by its two ends, loosely in their hands—the actual distance between these persons in a straight line is twenty feet—one of these persons raises his arms, and, by a strong impulse, shakes the cloth, while the other end is held by the other person firmly, a wave of the cloth is formed, and runs through its entire length, at the extremity of which it is lost. This is called undulation, or wave-making. The cloth rises and falls in the wave, which runs through twenty-five feet, its whole length. The distance traveled by the wave is twenty-five feet, being five feet more than the distance between the two persons holding the table-cloth. Should the table-cloth be stretched to its full length, no wave could be produced.

Now, let us apply this example to the sun and the earth. The luminous ether, as the intervening space between these two orbs is called, is ninety-two millions of miles in length; and, to admit of its undulation, must be very loose in its consistency. We may safely infer that such undulations as would be required for the transmission of light from the sun to the earth, would increase the actual distance traveled by the light in its undulations fully ten millions of miles, making the traveled space between the sun and earth to be one hundred and two millions of miles instead of ninety-two millions of miles, the measured distance. Now, the greatest velocity known is that of light,* which is 186,000 miles per second. We do no injustice to Divine Wisdom when we suppose that this extreme velocity has been imparted to light, in order that it should pass through space without interruption, and that it should reach its destination in the shortest possible space of time—in other words, that it should go directly to its object in right lines, without any deviation, up or down, or laterally, which would only retard its progress. Hence we reject entirely the undulatory theory of light, as enunciated at the present time. If the laws of light are not comprehended by scientists, it furnishes no excuse for resort to absurdities in the effort to explain them. While light, in traversing inter-stellar and inter-planetary spaces, is thought to be confined to rectilinear directions, there is nothing incompatible with this idea when it is brought within the influences of our atmosphere, by which its refrangibility, its reflection, its polarization, and its power to develop electricity, magnetism, and heat are manifested, and its more speedy diffusion through our atmosphere, by these disturbing influences, may furnish a reason for its attributes here, which would have no application in its passage through inter-stellar or inter-planetary spaces.

"Light diminishes in force or intensity in proportion as it recedes from its source. This diminution is *in direct ratio to the square of the distance.* Thus, the quantities of light at distances 2, 3, 4, etc., will be 4, 9, 16, etc., times less than at distance 1. Light requires eight minutes thirteen seconds to arrive from the sun to the earth. It travels $11\frac{1}{4}$ miles in $\frac{1}{1800}$ of a second, or 186,000 miles per second. It travels always in a straight line.

"Light added to light, by interference, produces darkness. The movement of such rays neutralize each other, and the light ceases to cast any lustre.

"Of the thousand rays of variegated shade and refrangibility

* Excepting that of electricity, which is 288,000 miles per second.

which compose colourless (or white) light, those only neutralize each other which possess co-ordinate colour and refrangibility. Thus a red ray cannot obliterate a green ray. Two white lights cross each other at a given point, and one time the red ray alone will disappear, and the point of intersection will become green—green being white minus red."

Let us see what can be made of the fragmentary knowledge of light that we have so far attained. The white light of the sun is composed of seven primary rays, all differing in colour from each other. The first analysis of this white sunlight was displayed to mankind in the rainbow, whose magnificent beauty was admired with stupid wonder, without the faintest conception on the part of the beholder of what it meant. After a lapse of ages of time, Sir Isaac Newton, with a glass prism, separated the rays of a sunbeam, and developed the primary colours which, in their association, had formed the white light of the sun. He reunited these primary rays, and thus, by synthesis as well as analysis, he proved the composite character of sunlight.

Now, astronomers have shown that the planets and asteroids of our planetary system each emit a colour peculiar to itself: Mercury, a pale rosy light; Mars, a reddish tint; Venus, a silvery-white colour, with occasional streaks of pale blue light; Jupiter gives out a pale yellow light; Saturn, a pale bluish tint, while its rings are gorgeous with a white, silvery colour; the Moon gives out a yellowish hue; Pallas shines with a yellowish light; Juno is a reddish star; Vesta has a ruddy tinge, sometimes of a pale yellowish hue; the Earth emits a red colour. "Another remarkable feature of these star systems, and perhaps the most brilliant and intrinsically beautiful phenomenon of astronomy, is the resplendent and gemlike variety of colours by which the binary, ternary and other multiple systems are characterized. Here all the colours and intermediate tints of the spectrum are to be met with, manifested with the richest intensity and the most vivid and distinctive strength and fulness ef hue. Thus in γ, Andromeda, we have a ternary combination, the brighter star being a rich and full orange, and the two fainter stars green. In α, Cassiopeiæ, we have a bright blue and a sea green star, β, Cygni, is a pair of stars, yellow and sapphire. α, Ceti, is a very fine orange star with a blue companion. * * *

"In a celebrated cluster of stars, near x of the Southern Cross, there are about one hundred small stars of different colours, from the various reds to all the tints of green, blue and bluish-green, so crowded together, that they appear in the larger telescopes like a piece of magnificent celestial jewelry, studded and flashing in the most superb splendour with the richest and most brilliant gem-light." * These colours are primary. What becomes of all these primary rays of light unless they are used to compose the white light of our sun, and of all the fixed stars or suns that illuminate the firmament? Whatever sunlight, therefore, has fallen upon these planets has been decomposed; six out of the seven primary rays thereof have been absorbed for the use of the planet, and the remaining primary has been emitted by the planet, and sent to the sun to associate in his photosphere with the different primary rays sent to him from other planets, to form anew the white sunlight, which by him is to be diffused throughout the planetary and stellar world.

Now we must not suppose that the orbs composing our diminutive solar system have furnished, or can furnish, to the sun a sufficient quantity of their respective primary rays of light to supply that luminary with the amount of elementary light which it is his function to combine and to furnish to the universe. We must remember that, from the great depths of the infinite expanse, elementary light comes up from every star, nebula, or meteor, seeking its complementary element in the photosphere of the sun, there to be associated as white light, and thence to be reflected from the gray covering of the sun, as a mirror, to all the orbs of creation. This circulation of light, this absorption by the stars and planets of such of the primary rays of light as they need for their own support, and the emission, severally, of their own peculiar rays, to be reassembled again in the various photospheres of the infinite number of suns that stud the firmament, and to be again diffused, according to the plan of creation, in endless succession, present an image of the wisdom, the beneficence and power of the Creator, that fills the mind with awe, and teaches man the utter insignificance of his being.

Our sun is simply a huge reflector of light. The gray covering of his nucleus or body is represented in our mirrors by the metallic covering which we place on the backs of our glasses. These transparent glasses are typified by the translucent photosphere of the sun, and the associated primary rays of light from every luminous object in the universe, mingling together, and reflected from this gray covering of the sun, furnish the white sunlight that illuminates the world.

---

*J. A. S. Rollwyn's Astronomy.

Heat destroys gravitation. Even our astronomers, in asserting that the luminous matter in the photosphere of the sun is shown by the spectroscope to be composed largely of incandescent metallic gases, the bases of which are among the heaviest matter in the crust of our earth, commit the inconsistency of supposing that these heavy incandescent metallic vapours or gases are supported by a photosphere of much greater specific gravity, as well as density, than these heavy gases themselves; otherwise these metallic gases could not float in the photosphere. Some of these astronomers go so far as to suppose that the body or nucleus of the sun itself is gaseous, and that the density of the sun is much less than the densities of the incandescent metallic vapours which they suppose to float in its photosphere. Now, if these incandescent metallic gases are heavier than the material composing the sun itself, it is clear that the gravitation, according to Newton, of these heavy metallic incandescent vapours is not towards the centre of the sun; and if not to him, where do they gravitate? We know what the specific gravities or densities of many of the metals on the surface of the earth are, whose incandescent vapours, as revealed by the spectoscope, are supposed to exist in the photosphere of the sun, and astronomers have calculated that the attraction of gravitation to the sun in its photosphere would be twenty-eight times as great as the gravitation in the earth's atmosphere to the earth of bodies of similar weight.

If, therefore, we suppose that these metallic incandescent vapours in the sun's photosphere to be twenty-eight times heavier than they would be in the earth's atmosphere; and if they never fall to the body of the sun, it must follow that what is called gravitation in the photosphere of the sun cannot exist, and the whole theory of Newton, of centripetal and centrifugal forces, has no substantial existence. We know that in our own planet heat destroys gravitation, as the volcanic action in the interior of the earth, upheaving islands, mountain ranges, and even continents, abundantly proves

The mean density of the earth is about five times greater than that of water—actually 5.44 times. Water, therefore, rests on the surface of the earth—penetrates its crust till it encounters the heat radiated from the interior of the earth, where its further descent below the surface is arrested, then it is converted into steam by the heat it has absorbed, and it is driven upwards into the atmosphere, heaving up the most solid and heavy materials of the crust of the earth, that lie above the direction it may take. This expansion of water into steam by heat in the crust of the earth, produced by the repellent affinity of the homogeneous electricity associated with it, is one of the forces of volcanic action, which are continually changing the forms of the outer surface of the earth's crust. The density or specific gravity of the sun is 0.25136 (or nearly one-fourth of that of the earth). In other words, taken in equal volumes, the weight of the matter which composes the sun is scarcely more than one-fourth of the weight which composes our globe. Compared to water, the density of the sun is 1.367; that of water being 1.

Now, if what our astronomers tell us of the inconceivably high temperature of the sun be true, there can be no gravitation towards its centre from its photosphere, its chromosphere, or any of its possible envelopes, the heat expanding, rarefying and driving off all such material substances. Heat disintegrates solids, separates their molecules, destroys their densities, and consequently is opposed to gravitation, which is the attraction of densities. Alas! for poor Sir Isaac Newton and his grand theory of centripetal and centrifugal forces! A ray of light passing through a narrow chink, and through a glass prism, has done the business. The incandescent metallic gases and the transcendent intense heat of the sun which has vapourized these metals (the supposed discovery by the narrow chink and the prism), have demolished Newton and his erratic fancies. *Sic transit gloria mundi!*

According to Professor Tyndall, "gravitation consists of an attraction of every particle of matter for every other particle—planets and moons are supposed to be held in their orbits by this attraction."

"The earth is supposed to attract to its centre all the bodies upon its surface by what Newton termed centripetal force, and when one of them falls, it is always towards the earth's centre. This force is said to be resident in all the bodies of nature. It exerts its influence upon the largest masses as well as upon the most minute particles of matter. This it is which gives harmony to the universe, and explains the formation of bodies of all kinds."

Newton held that "Bodies exercise attraction in direct ratio to their mass, and that this law was of universal application."

Let us examine this.

The circulation of the blood in animals is not affected by gravitation, nor are any of the secretions of the animal body. The development in growth of animals is upwards, opposed to gravitation, and totally unaffected by gravitation. The movements of animals in the performance of their varied functions have no reference to gravitation. So also in the vegetable world; the sap of plants rises from the roots, is distributed through the branches, and enlarges their size irrespective of gravitation; the trunk of the tree ascends into the atmosphere and extends its huge limbs laterally, as if gravitation had no existence. The smoke from combustion, the exhalations from the earth, and the evaporation of water, all of them material substances, are in opposition to gravitation.

Light, electricity, magnetism and heat, the vital forces of the universe, all treat gravitation with great contempt. The atmosphere surrounds and envelopes the earth. It has what is called gravity or weight, but it is not subject to what is called the law of gravitation, since when its lower strata become warmed, they ascend into the upper part of the atmosphere, and do not descend or fall to the earth, as having weight they should do; thus a difference in the relative weights of the same substance, in one condition or another, removes that substance from the influence of gravitation. The vapours or clouds in the atmosphere, which are heavier than air, float in many directions, and do not fall to the earth. A piece of iron will float upon a fused mass of iron, instead of passing through it to the bottom. The inertia of matter is opposed to gravitation. Form, which is a force, and is the resultant of the forces that have produced it, is antagonistic to gravitation, which we illustrate with this example: suppose we have a cube of soft iron, weighing five pounds; let it be held by the hand over a pool of water; release it from the hand, the iron falls directly to the bottom of the pool; our philosophers would say it fell by gravitation.

Now, take that cube of iron, roll it out into a sheet of iron one-sixteenth of an inch in thickness, and again place it over the water horizontally; release your hold upon it; it sinks immediately to the bottom of the pool. Philosophy says, by gravitation. Recover it, and holding its edge vertically over the water, again withdraw your hand; it descends at once to the bottom. Still by gravitation. Now, again take it from the pool, bend its edges up some six inches around it, in the form of a dish; then place its bottom on the surface of the water, release your hold, and lo! it does not sink to the bottom of the pool, but it floats upon the surface of it! It is no longer drawn to the bottom of the pool by gravitation, although what we call its weight is unchanged. It still weighs five pounds. Why does it not sink as before? It is arrested by its form, which is antagonistic to what is called gravitation. Gravitation, therefore, is not universal. It does not always attract matter to matter, in proportion to its mass. What then is the repellent force which prevents this iron dish from sinking? It is magnetism. The water is magnetic, a condition produced by the electricity, whose opposite polarities in the oxygen and hydrogen meeting in conjunction, converted those gases, by the combustion of the hydrogen gas in the oxygen gas, into the liquid state of water, and rendering the water at the same time magnetic. The iron dish, in contact with the water by its horizontal bottom, and having vertical sides, became magnetic by induction from the water—the water and the iron presenting the same magnetic poles to each other, mutually repelled each other, and the flotation of the iron dish was the result.

Flotation, heretofore attributed to the lightness of the floating body compared with the weight of the liquid in which it floated, is due to magnetic repulsion, and not to gravitation. Now let us look at the condition of this water when it has changed its character by crystalizing into flakes of snow, of whatever diversity of form, or of hail, or of surface or dense ice. These forms of water at temperatures below 32° of Fahrenheit, are all magnets, and their minutest atoms are all magnets, also; each endowed with its two poles, one at either extremity of the atom, and each with opposite attributes.

The commerce of the world, therefore, is sustained on its oceans by the repellent force of magnetism; while the mariner directs his course over their trackless wastes, in darkness and in storm, guided by that opposite quality of the magnet which attracts it to the poles of the earth.

Now, when water, owing its form, whether liquid or frozen, to magnetism, is exposed to heat, and converted into steam, its magnetic qualities are driven off by the heat, and are replaced by electricity, which is the force that rends the strongest fabrics of human skill to pieces, and scatters death and desolation in every direction. The electricity of steam is of one kind, and is repellent of itself; and its effort to escape from itself and to unite with the opposite electricity of the atmosphere is so violent and so powerful that it furnishes to man one of the greatest forces with which he is acquainted.

The forked flashes of lightning, seen above volcanoes in eruption, are merely the results of the conjunction of the positive electricity of the heated air, steam and lava thrown out of the volcano by violent interior forces, with the negative electricity of the atmosphere above and around the volcano.

Rotary motion of an object is antagonistic to magnetism, by the production of friction with the atmosphere by the revolving object. This friction evolves electricity, which, uniting with the opposite electricity of the revolving object, produces heat that expands and disintegrates its molecules, separating them, and removing the magnetism.

As the heat of the sun (if it has any) cannot pass downwards through ninety-two millions of miles of ether with a temperature of —142° of centigrade thermometer, so the heat radiated from the interior of the earth, or produced on its surface,,or, in its lower strata of atmosphere, cannot penetrate upwards through the canopy of cold which surrounds the earth at various altitudes from the snow line of 15 000 feet above the equator, 6000 feet at 45° of north or south latitude, and at the level of the earth at 60° of north latitude.

Let us admire the ineffable wisdom of the Creator who, by a barrier of ice in the Arctic and Antarctic regions, confines the internal heat between them and the equator, and the superficial heat of the earth below the region of perpetual snow in the atmosphere, for the uses intended by Him of the planet and its productions

Newton's theory of centripetal and centrifugal attractions and repulsions is fallacious. There can be no rotation on the centre of a sphere or spheroid, though there may be at the extremities of any of its diameters or axes. What is called centrifugal force is merely the repulsion from the axis of rotation and not from the centre. So centripetal force is merely axial attraction. Any force is the resultant of the forces which produce it. If there was, therefore, such a force as centripetal in a sphere or spheroid, the opposing forces acting from the ends of the diameters would neutralize each other, and an immense heat would result at the centre, which heat would destroy the very forces which had produced it, and would prevent their continuance.

When we consider the repellent forces of the interior of the earth, such as heat and electricity, upheaving by volcanic action immense masses of islands and continents, changing in many places the configuration of the land and the sea, we cannot for a moment accept the theory of centripetal attraction or gravitation.

The mean density of the earth is said to be about five times greater than that of water. If this be so, why does not this great density or mass of matter bring down the clouds by centripetal attraction or gravitation instantly to the earth? Why does the atmosphere, still less dense than the clouds, remain above the earth, when according to the laws of gravitation it should be precipitated upon it? and why should the upper strata of the atmosphere be more attenuated and thin than the lower strata, which besides their own weight have the additional weight of the upper strata upon them?

There are no centripetal or centrifugal forces, as Newton supposed. In the rapid rotation of a sphere or cylinder on its axis, the outer surface, by its friction with the atmosphere, evolves electricity, which, in conjunction with the electricity of the atmosphere, produces heat, which insinuating itself among the molecules of the rotating body, expands them and, if the velocity of the rotation is sufficient, this heat loosens their mutual cohesion, and electricity being at the same time imparted to these molecules associated with the heat, they are attracted thereby to the opposite electricity of the atmosphere, and the rotating body is separated into fragments with great violence, as the molecules of the mass, having the same electricity, repel each other while they are attracted to the opposite electricity of the outer atmosphere.

This is the explanation of the bursting of millstones, grindstones and other revolving bodies at great speed, as well as of meteors, shooting stars and comets, heretofore attributed to centrifugal force. Now, what is there to attract at the centre of anything or to repel therefrom. The centre is an imaginary point, having neither length, breadth nor thickness, absolutely without dimensions, and consequently without matter—how therefore can it be invested with force of any kind?

There can be no rotation on the centre of any sphere,

cylinder, or cone, or other solid or hollow body, as the forces requisite to produce the motion, would be antagonistic, and would destroy it, as the attempt might be made—conceive for a moment, that while the earth is revolving on its axis from west to east, you should apply an equal force to make it revolve also from north to south, the rotation would then be from northwest to southeast—now apply equal intermediate forces between northwest and west, and northwest and north, and so on till you have equal forces for every degree of the hemisphere, and equal opposite forces from the other hemisphere. This would be equivalent to centripetal force or attraction, and as these opposing forces would be equal, rotation would cease, the body would remain at rest, and centripetal force or attraction would not exist, consequently there is neither centripetal nor centrifugal force, and we must look therefore to other forces to explain the motions of the planetary and stellar worlds.

It is to Oërsted, the celebrated chemist and physicist of Denmark, that we owe the discovery that currents of electricity passing over a conjunctive wire, from one pole of the Voltaic pile to the opposite pole, produce magnetism. The meeting of these opposite electricities, he has termed an "*electrical conflict.*" I should prefer to call it an electrical embrace, as it more resembles the ardour of lovers, in its attraction, than an attack by force or violence. From his experiments he concluded that the electric conflict is not inclosed in the conducting wire, but that it has around it quite an extensive sphere of activity, and that it acts by a vortical or whirling movement.

A few weeks after the announcement of Oërsted's discovery, Ampère, by his experiments, discovered that two parallel conjunctive wires, from opposite poles of a Voltaic pile, attract each other, when electricity traverses them in the same direction; and that they repel each other if the electric currents move in opposite directions. The sequel of Ampère's labours showed that the reciprocal action of the elements of two currents is exerted in conformity with the line which unites their centres; that it depends on the mutual inclination of these elements, and that it varies in intensity in the inverse ratio of the squares of the distances. Ampère finally succeeded in establishing that a conjunctive wire wound into a helix or spiral curved line, with very close spires, is sensitive to the magnetic action of the earth. For many weeks there was to be seen in his cabinet a conjunctive wire of platina, whose position was determined by the action of the terrestrial globe. Ampère, by constructing a galvanic compass, had shown that the forces which act in the magnetic needle are electric currents, and by his learned calculations on the the reciprocal action of these currents, he accounted for all the actions which the conjunctive wire of the pile exerts, in the experiment of Oersted, on the magnetic needle.

M. Arago, the eminent French astronomer, associated with Ampère in some of his experiments, says: "I coiled copper wire for a length of two inches, from right to left, into a helix; then an equal length of wire in the same manner, from left to right; and lastly, a similar quantity again from right to left. These three helices were separated from each other by rectilinear portions of the same wire.

"One and the same steel cylinder of a suitable length and of rather more than .04 of an inch diameter, and enclosed in a glass tube, was inserted in the three helices at once. The galvanic current, in passing along the coils of these different helices, magnetized the corresponding portions of the steel cylinder, as if they had been detached and separate from each other; for I remarked that at one of the extremities there was a north pole, at two inches distance a south pole, farther on a second south pole followed by a north pole; lastly, a third north pole, and two inches farther on, or at the other extremity of the cylinder, a south pole." Thus, by this method, the number of these intermediate poles, which physicists have denominated consecutive points, could be multiplied at pleasure. M. Arago also observed, that "if the intervals comprised between the consecutive helices are small, the parts of the steel wire or cylinder, corresponding to those intervals, will themselves be magnetized as if the movement of rotation impressed on the magnetic fluid, according to Ampère's idea, by the influence of a helix, was continued beyond the extreme spires of the coil."

As the conjunction of opposite electricities, according to these authorities, develops magnetism; and as tornadoes, hurricanes, cyclones, and other atmospheric disturbances move in spiral curves from their respective points of departure till their terminations, and as, according to Ampère and Arago, currents of electricity passed through spiral cylindrical coils of wire develop magnetism, we see here the sources of the supply of magnetism to our planet, its atmosphere, and the objects upon or in them. This magnetism, so developed, is absorbed by every object in nature. Being an imponderable, its presence cannot always be discerned or detected; but it resides in a latent form everywhere, till it is evolved by the opposite attraction or repulsion of some object approached to it which is also magnetic.

In many parts of the world springs of water exist in which a great degree of magnetic power is manifested. In the state of Michigan there are such springs, in which, if penknives, or small pieces of iron, or steel, should be immersed for a few minutes, they would become highly magnetic. These springs are visited and bathed in every year by thousands of persons for the highly curative influences over diseases that they exert.

There is no magnetism in the earth under the equatorial regions, owing to the heat of the interior of the central parts of the planet, which destroys magnetism. This is proved by the magnetic needle losing its dip under the equator. I think, also, it will be shown that the magnetic needle has no dip over the Gulf stream, as under that stream the interior heat of the earth has a flue extending far into the Arctic regions, through which the Gulf stream is warmed, and magnetism in the earth about the flue destroyed; the same will be found to be true, also, of the Japanese current that runs through Behring's strait to the Arctic regions; and of all other warm currents of water in the oceans. The evaporation of the warm waters of the Gulf stream and of the Japanese current develops electricity, which, being positive as the waters thereof themselves also are, they are both attracted by the negative electricity of the waters of the Arctic ocean; and those currents flow in that direction. It will be found that terrestrial magnetism is irregularly distributed in the crust of the earth, and the magnetism of the Northern Hemisphere being attracted to the South Pole, while that in the Southern Hemisphere being attracted to the North Pole, these opposite attractions have increased the equatorial diameter of the earth twenty-six miles more than the polar diameter; and the earth's crust under the equator having been thickened by the addition of so much material taken from other parts of the sphere, it follows as highly probable that basins filled with seas have resulted at the poles of the earth, and that oceanic currents from the North and South Poles, respectively, are produced by the rotation of the earth on its axis, throwing off the surplus of accumulated water at the poles, and thus the circulation of water in oceans and seas is produced, in spiral curves from the polar basins.

. I have, in the former editions of this work, suggested that the rotation of the earth on its axis is the result of electrical forces within it, excited by the juxtaposition of the materials of various kinds forming its composition, and having opposite electrical polarities.

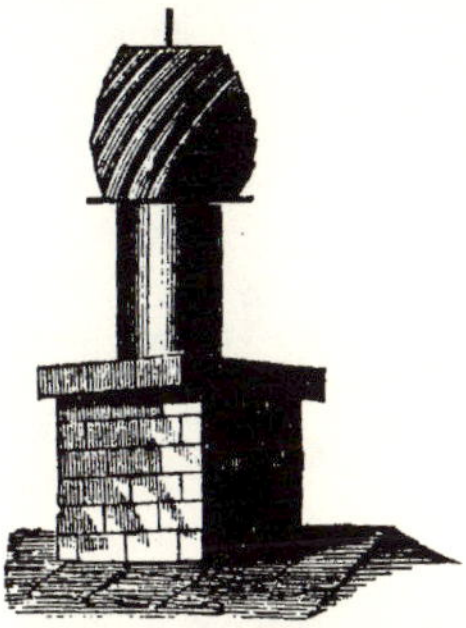

I have an illustration at hand to prove this. A neighbour of mine recently erected in the rear of his house a one-storied dining-room, in which was a chimney which projected some three feet above the roof of the building—which was 12 feet above the ground—on the top of the chimney he placed a sheet-iron cowl in the form of a truncated hollow ellipsoid with spiral flanges from top to bottom of the cowl. When there is no fire in the chimney the cowl is at rest, when a fire is kindled, as the air in the chimney becomes heated and, accompanied by its positive electricity, rises to the top, it meets with resistance in the flanges of the cowl, which only begin to turn when the gathering positive electricity of the warm air attracted by the greater negative electricity of the outer atmosphere forces its way through the openings and along the surface of the metalic cowl and sets it in motion, and according as the combustion is more active so is the rotation of the cowl on its axis the more rapid, and the draught of the chimney is so increased that finally the flanges of the cowl can no longer be distinguished in their rotation.

So in the interior of the earth the intense positive electricity evolved there, in conjunction with the negative electricity also there in great quantities, produces enormous heat, which fusing metals and disengaging gases of great volume and expansive power, forces them against the irregular surfaces of the interior of the crust of the earth, and sets the ball in its rotary motion on its axis.

Similar causes produce like effects in the interior of the sun and of all the planets, giving them all the rotation on their respective axes that we know they have. With the electricity thus evolved and escaping as it is formed at their respective poles, currents of magnetism are evolved at right angles to the currents of electricity and cause the revolutions on their axes to be from west to east.

There is no necessity, therefore, for our astronomers to suppose that the Almighty has created the sun to be an incandescent body, whose combustion is to be fed by half a world to illuminate the remainder. The sun, in fact, is probably only a huge reflector or mirror, receiving the rays of light from every orb, which rays themselves are of various tints, as every planet and star has a colour peculiar to itself, and the groupings of these primary colours in the sun, and their reflections from him constitute the white light that we call sunlight. This explanation is in harmony with our ideas of the Divine economy, which never wastes any of its material. The sun is a great magnet, and regulates and controls by magnetism and not by gravitation all the planets of his system, which, consequently, are severally all magnets. The system is held in its place and conforms in its movements by its magnetism to the movements of all the orbs which exist in space.

As these planets are all magnets, they can have no other heat than their own internal heat, which is simply sufficient to produce their respective rotations on their several axes, as heat in intensity destroys magnetism.

The reversal of the tails of comets in their approach to the sun and departure from him, is due to the attraction and repulsion respectively of their magnetic poles—by induction from the greater magnetism of the sun itself.

Winds are simply currents of electrified air, repelled from their points of departure by air similarly electrified, and attracted in their various directions by air at rest or in motion, as it may be, with opposite electricities. These repellent and attractive electricities acting on a strong current of air, cause it to be deflected from its rectilinear direction, and to assume a spiral curve in its course, continually contracting towards its centre, till the opposing electricities equalize each other, when the electrical equilibrium is restored, and a calm ensues. During the continuance of the movements of the oppositely electrified currents of air in these spiral curves, magnetism is developed, and this is the source of magnetism in the atmosphere.

Magnetism in the crust of the earth is likewise developed

there by the conjuuction of opposite electrical currents circulating continually through it. This magnetism permeates through its various molecules, supplying them with magnetic attraction and repulsion, and thus matter, from its susceptibility of becoming magnetized, assumes the power of attraction attributed to gravitation.

Having thus shown the source from which atmospheric as well as terrestrial magnetism is derived, we proceed to mention some of its attributes.

The term magnetism, which is applied to the science that describes the modes and properties of a remarkable force possessing attractive and repellent qualities, is derived from a magnetic iron ore, that was first noticed near Magnesia, and hence was named by the ancient Greeks, *Magnes.* It had the peculiar property of attracting iron. This force is not confined to the mineral, but seems to pervade all nature. It is produced by the meeting of currents of opposite electricities in the crust of the earth and in our atmosphere. Its existence in the fixed stars, in the infinite number of orbs, in the firmament, in the nebulæ, comets, meteors, &c., may be attributed to a similar origin. The primary rays of light from these illuminated orbs, of greatly diversified colours, passing with almost incredible velocity from them to our sun, through interstellar and interplanetary spaces whose temperature is inconceivably low, and consequently associated with negative electricity, developing as they pass through this attenuated ether, which fills these spaces, by friction therewith, negative electricity, may be supposed to enter the photosphere of the sun charged with negative electricity. This negative electricity being homogeneous, of immense volume, and great intensity, repels these commingled primary rays of light, by reflection from the body of the sun on their impact with it, with the enormous velocity which belongs to light. The mixture of these primary rays of various colours produces the white light of the sun, or, as we call it, sunlight. This sunlight, negatively electrified, driven with this immense speed to the most distant orbs of creation, encounters in their atmosphere, when such exist, and by impact with the bodies of these orbs themselves, which have each a greater density than has the ether through which it had passed, great resistance. This impact produces friction, and friction electricity.

The friction of matter having a temperature above 32° of

Fahrenheit evolves positive electricity, while that of matter whose temperature is below 32° of Fahrenheit evolves negative electricity. When two blocks of ice are rubbed together they adhere by their contiguous surfaces with a force greater than that by which the molecules of either block of ice are held together, and a fracture of the ice will occur anywhere in the blocks before it will at their junction. A notable illustration of the friction of matter, below 32° of Fahrenheit, producing cold and its associate negative electricity, is furnished every day in the manufacture of iced creams and juices of fruits. The cylinder containing the material to be frozen is placed in another vessel, surrounded by a freezing mixture of broken ice and common salt; by turning this cylinder rapidly in this mixture friction is produced, which, in abstracting the heat from the cream or juices of fruits to be frozen, reduces their temperature, and the cold of the freezing mixture, with its negative electricity, is transferred to the cream or juices of fruits.

We may infer an analogy between the composition of these distant orbs of the firmament and that of our own planet, and that an opposite electricity to that of sunlight exists in them. The conjunction of these opposing electricities develops magnetism, which at once seizes upon the matter of which such orbs are composed and imparts to it the attractive and repellent qualities that it possesses. The orb assumes the form of an oblate spheroid or an ellipsoid, with its equatorial diameter longer than its polar diameter, thickened at its equator and flattened at its poles. This form imposes on it an elliptical orbit in which it revolves around its local attraction. This form in the planets and probably the fixed stars, as in the earth, is derived from the opposite attractions and repulsions of matter in their different hemispheres—that in their northern hemisphere being attracted to the south pole, and that in the southern hemisphere being oppositely attracted to the north pole—and thus meeting at their respective equators, where these opposite attractions neutralize each other, they become thickened there at the expense of the matter at their poles respectively. The force which drives the sunlight from our sun, after its reflection from its body, is probably negative electricity, for we cannot conceive of any other force adequate to produce such an effect.

It is this force of magnetism of which Newton in his day had some slight knowledge, but not comprehending it as it exists. he assigned such of its qualities as he had discovered erroneously to matter, and gave it the name of gravitation, as if a planet, if such could be made, of cotton, rice, tobacco, butter, cheese and molasses, would revolve upon its axis from its own weight and travel in an orbit around the sun.

This force magnetises all things, imparting to them its attractions and repulsions, and thus regulates and controls the movements throughout the universe.

Let us notice some of the pecularities of this force. "Some iron ores are natural magnets; steel rods, straight, or curved like horseshoes, to which magnetism has been imparted, as also steel needles similarly treated, are artificial magnets. The magnetic force is greatest at the ends of the rods or needles, attracting there steel or iron filings, but diminishing in power as the distance from the extremities is increased, and ceasing altogether midway between their ends. The extremities of the rods or needles are called its poles; midway between them, where the force ceases, is called their magnetic equator. A light needle magnetised, such as is used in the mariner's compass, properly balanced and suspended by its centre is called a magnetic needle. When not restrained it ranges itself nearly parallel to a line joining the north and south poles of the earth, one end of the needle pointing to the north, the other end directed to the south pole. Turned from its direction and then released, it resumes again its natural position of pointing north and south. These ends or poles of a magnet are respectively attached to the poles of the earth to which they point, and are repelled from the opposite poles reciprocally. In two magnets the corresponding poles, if approached to each other, would each repel the other and attract the opposite pole of the other magnet." It is to this attribute of the magnet that the earth owes its form of an oblate spheroid. The earth being a magnet, the materials composing its crust in the northern hemisphere have been attracted towards the sonth pole, and the matter in the earth's crust in the southern hemisphere, being also magnetic, have been attracted towards the north pole. These forces being equal and having ceased at the equator, the matters brought by them respectively from their several hemispheres have been accumulated and deposited in the equatorial regions of the earth, which mass of matters has so much increased the equatorial diameter of the earth that it exceeds the polar diameter in length 26 miles. It is probable that the material thus removed from the poles of the earth to its equator, have so hollowed out the crust of the earth at the poles into basins that seas have been formed in them, which have been filled with water from the Pacific ocean through Behring's straits, and Atlantic ocean by the Gulf Stream. As the planets are all doubtless formed upon the same principle as those on which the earth is established, and as we know that similar differences exist between the equatorial and polar diameters of these orbs to the extent of 25 miles in Mars, 6000 miles in Jupiter, and 7500 miles in Saturn, we may reasonably infer that magnetic attraction and repulsion have increased their equatorial diameters at the expense of their polar diameters in the proportions mentioned, and that like the earth they are all magnets, and owe their axial and orbitual rotations to magnetism, and not to gravitation. In this increase of matter in the equatorial regions of these planets of our system, we have the most conclusive evidence that the attraction of matter in these orbs is to their respective equators, and not to their respective centres as Newton supposed.

When we regard these immense differences in the equatorial and polar diameters of the planets, Jupiter and Saturn—that of Jupiter being 6000 miles, and that of Saturn 7500 miles, we begin to comprehend, in a slight degree, the idea of the Creator in placing these planets at such immensely great distances from the sun, while He invests them with a magnetism so transcendantly powerful in its attractions and repulsions, that their revolutions around the sun are performed with a marvelous certainty and exactitude. The law of magnetic attraction and repulsion between objects being inversely as the square of the distance, those distant orbs must have a propelling or repellent power at their greatest distances from the sun of almost infinite magnitude, to bring them within the attractive power of the sun, so as to pass over such immense spaces in their allotted times. It is the repellent power of magnetism that returns them towards the sun.

"Similiar poles of a magnet repel, and contrary poles attract one another; magnetic poles always occur in pairs. If a magnet be broken into many pieces, each fragment is found to have its north and south poles.

"Magnetic attraction and repulsion vary inversely as the square of the distance between the magnet and the body attracted or repelled.

"If in two magnets of equal strength, the north pole of one

of them be placed in contact with the south pole of the other magnet, all attractive force will disappear. Remove the contact, and the magnetic force is restored in each of the magnets.

"If a pole of a permanent magnet is placed near to the end of a bar of soft iron, this bar will be magnetized by *induction*, the end of the soft bar next to the pole of the magnet having there an opposite pole to that of the magnets, while at the other end of the iron bar will be found a contrary magnetic pole. Magnetization by *induction*, may be effected through a plate of glass, wood, metal, &c., without detriment. This condition vanishes as soon as the magnet is withdrawn.

"Besides iron and steel, nickel, cobalt, manganese, chromium, platinum, oxygen gas and many other substances, suffer attraction by a magnet. Heat powerfully influences magnetism. A magnet if heated to redness, loses all its magnetism, and a red hot ball is not atttracted by a magnet.

"Every magnetic substance has its limit of temperature; thus cobalt does not cease to be attracted at a white heat; iron ceases to be attracted at a red heat; chromium just below a red heat; nickel at 350° Fahrenheit; and manganese is not attracted on a warm summer day. Hence it is probable that certain substances which do not appear, under ordinary circumstances, to be attracted by a magnet would be attracted if their temperature was reduced to a sufficiently low degree.

"A magnetic needle tends to set itself in a line with the poles of the earth, and if moved from this position returns to it, as if it was in the presence of another magnet. This is due to the magnetism of the earth—in fact, the earth is a huge magnet, the poles and equator of which do not coincide with the geographical poles and equator.

"The magnetic meridian of a place is a vertical plane which passes through the two poles of a horizontally suspended magnetic needle at this place, and which being continued in both directions will, of course, pass through the magnetic poles of the earth. The magnetic meridian of a place will not coincide with its geographical meridian, and the angle formed by the two meridians is called the magnetic *deviation*, *variation* or *declination*, at this place.

"The variation of the needle does not always remain the same. In the year 1580 (the first year in which accurate observations were made) the north end of the needle deviated 11° 15′ to the east of the true north in London. In 1622 the deviation was 6° east of the north, and in 1660 the magnetic north pole coincided with the geographical north pole. In 1692 it had passed to 6° west of north. In 1765 it was 20° west; and in 1818 it attained its maximum westerly deviation—24° 41′. It is now returning to the north. In 1850 the westerly deviation was 22° 30′; and in October, 1871, the deviation observed at the Kerr Observatory was 20° 18′ 7″. This is the *secular* variation of the magnetic needle. A delicately suspended magnet may be observed to undergo an annual, daily, and even hourly variation.

"If a steel needle be accurately balanced about a horizontal centre, and be there magnetized, it will no longer be in horizontal equilibrium. In London the north end of the needle will dip down, forming an angle of more than 60°, with a horizontal plane. The angle which a magnetic needle, capable of vertical movement, (*dipping needle*,) makes with a horizontal plane is called the angle of *inclination* or *dip*. The vertical plane in which the needle moves must coincide with the magnetic meridian of the place.

"The dip varies in different parts of the world. If we convey a dipping needle north of London the dip increases; if, on the other hand, we go south of London the dip diminishes; at the magnetic equator there is no dip, the needle is perfectly horizontal; and south of the equator the south pole of the needle begins to dip, and the dip increases as we go further south. Thus the dip at Peru is 0°, at Lima 10° 30′, at the Cape of Good Hope 34°, and at Hudson's Bay between 89° and 90°.

"The *magnetic poles* of the earth are those points on the earth's surface at which a dipping needle assumes a vertical position. The north magnetic pole was discovered by Sir James Ross, in 1830. It is situated in longitude 96° 43′ west, latitude 79° north. The south magnetic pole, is as yet, unknown.

"The magnetic equator of the earth is a line connecting all those places on the earth's surface, at which there is no dip. It is an irregular closed circular line cutting the terrestrial equator at four points. The dip of a magnetic needle is subject to both secular and periodic changes. Thus in 1576 it was 71° 51′ in London; a hundred years later, it was 73°

30′, and in 1723, it reached a maximum of 74° 42′. In 1800, it had decreased to 70° 35′, and in October 1871, the dip registered at the Kew Observatory was 67° 56′ 3″. The dip also undergoes annual and daily changes.

"If a horizontally suspended magnetic needle be moved from its position of rest, it returns to it, passes it, and oscillates backwards and forwards across the final position of rest in the magnetic meridian of the place; in fact, it becomes a horizontal pendulum oscillating under the influence of the earth's magnetism. It has been proved that the intensity of the earth's magnetism, at any two places, is proportional to the square of the number of oscillations made by the same magnetic needle at these places.

"Various determinations of the intensity of the earth's magnetism prove that the force increases as we pass from the equator to the poles, as in an ordinary magnet. Thus if the intensity at Peru be taken as unity, the intensity in London will be represented by 1.369, and at Baffin's Bay by 1.707.

"All matter is affected by a powerful magnet, but while many substances (iron, nickel, manganese, oxygen gas, &c.,) are attracted, other substances (bismuth, copper, hydrogen, &c.,) are repelled by *both* poles of the magnet.

"If a small bar of iron or other attracted substance, be suspended between the poles of a magnet, the bar will set itself *axially*, that is with its length in a line joining the two poles. If on the other hand a bar of bismuth or other repelled substance be suspended in a like position, it will set itself *equatorially*, that is at right angles to a line joining the poles of the magnet, because as it is repelled by both poles, it will endeavor to keep as far away from them as possible. Such bodies are called *dia-magnetic*."

In Professor Tyndall's introduction to his "Researches on Dia-Magnetism," writing of Professor Faraday, he states, "That having laid hold of the fact of repulsion, he immediately expanded and multiplied it. He subjected bodies of the most various qualities to the action of his magnet; mineral salts, acids, alkalies, ethers, alcohols, aqueous solutions, glass, phosphorus, resins, oils, essences, vegetable and animal tissues, and found them all amenable to magnetic influence. No known solid or liquid proved insensible to the magnetic power. When developed in sufficient strength, all the tissues of the human body, the blood—though it contains iron—included, were proved to be dia-magnetic, so that if you could suspend a man between the poles of a magnet, his extremities would retreat from the poles, until his length became equatorial," that is to say, horizontally perpendicular to the magnetic meridian.

From the dip or inclination of the magnetic needle on various parts of the earth's surface—as magnetism is a dual force—we infer that one of its poles is attracted by the magnetism existing in the upper atmosphere, while the other is attracted to the magnetism in the crust of the earth beneath. At Peru the dip is 0°, owing probably to the heat in the interior of the earth under Peru, which is frequently manifested in the most violent earthquakes and volcanic action, and heat we know destroys magnetism. As the dip of the needle in either hemisphere increases from the magnetic equator toward the poles, it is obvious that the magnetism in the upper atmosphere, as well as in the crust of the earth, also increases in a like proportion, attributable doubtless to the increased cold, both of the upper atmosphere and the crust of the earth in high latitude, and as negative electricity and magnetism are both associated with extreme cold, we find herein an explanation of the dip of the magnetic needle.

In the attraction and repulsion of the magnetic needle, horizontally, at the magnetic equator towards the north and south poles of the earth, we have a dual horizontal force. In the deviation of the needle east or west of north or south, we have another dual force acting horizontally. In the class of subjects called dia-magnetic, which arrange themselves at right angles to the magnetic meridian, or equatorially as it is termed, we have another dual force acting horizontally. In the dip of the needle, which is nothing at the magnetic equator, but whose angle with the horizon increasing therefrom as we advance towards either pole till it reaches 90° or a quadrant of a circle, we find another dual force with one set of poles in the frozen crust of the earth, while an opposite set of poles is in the equally frozen regions of the arctic and antarctic upper atmosphere of our planet.

These forces, with electricity and heat, all developed by light and controlled by the omniscient wisdom of the Almighty, are the powers which regulate the motions of our planet and preserve it in its integrity.

We may well dispense, therefore, with the whole theory of centripetal and centrifugal forces, and of the attraction of matter by weight, which continually is being changed with the forms and positions it assumes, the same substance being at one time solid and fixed to the earth, then liquid and movable on its surface and again gaseous and floating in its atmosphere above it.

In connection with this subject of magnetism, it is curious to observe that in the animal and vegetable kingdoms the forms of their productions all conform, in a greater or lesser degree, to the typical forms of ellipsoids, or oblate spheroids, as manifested in the planets. Examine the forms of our trees. Vertical or horizontal sections, when they are in full leaf, would disclose curved lines, which, if tangential to the extremities of their leafy branches, would represent the elements of an ellipse—in some cases elongated, in others approaching nearly to the form of a circle. So with their leaves, however long and narrow they may be, the elemental character of the ellipse is apparent in them. The fruits they bear have all similar characteristics. The apple, the peach, the pear, the apricot, the nectarine, and indeed all the stone fruits, have shapes corresponding nearly to the ellipsoid. The nut-bearing trees, from the cocoa-nut through the walnuts, hickories, pecan nuts, chestnuts and beeches, all produce fruits which, in their outer forms, partake of the character of ellipsoids, or oblate spheroids. The coffee berry, the olive, the fig, the date, all correspond in their general forms to the same type. Among what are called vegetables, from the enormous melon, in all its varieties, through the pod-bearing plants, the cabbage, &c., the same type is visible. So in the roots and tubers; the turnip is an oblate spheroid, the potato commonly an ellipsoid, as are also the carrot and the parsnip. In the seeds of the family of grapes, as well as in their leaves, the same forms are found. The bunches of grapes, as well as their berries, are all of the same characteristic form. Take even the grasses —in which may be included the cereals. Their long and narrow leaves are all elliptical in form, though they may, in some cases, be pointed at their outer extremities. These long leaves assume the form of a semi-ellipse, in their curvature from the stem or branches, from which they grow, towards the ground. So it is with the long blades of maize or Indian corn, the sugar cane, and sorghum. The leaves, fruits and branches of trees, for the most part, have an inclination towards the earth, and are commonly pendant. Their tops are attracted upwards, and are frequently vertical. Why do their branches extend laterally and downwards, while their trunks and summits ascend vertically in the atmosphere? And why do their leaves and fruits hang downwards? Is it not because of their magnetic condition? Now, the leaves, fruits and branches of trees, which pursue horizontal, or slightly inclined directions, may be supposed to be dia-magnetic, and under the influence of the horizontal currents of magnetism that set equatorially to the magnetic meridian; while the trunks and summits, repelled by the magnetism of the earth, are attracted by the opposite magnetism of the upper atmosphere, and rise vertically. These two forces, varying in intensity, produce all the resultant directions which their branches assume in their development. Fruits of trees, being ellipsoidal in form, (which is the common form of simple magnets,) and generally pendant vertically, when they fall to the ground are attracted there by the superior magnetism of the earth, and remain on it by the same attraction, unless removed from it by a superior force.

If there is any truth in the story of Sir Isaac Newton having been led to the adoption of his theory of gravitation, and of centripetal and centrifugal forces, by the sight of an apple falling from its tree to the ground, it is to be lamented that he did not investigate the force which expanded the seed, caused its germination, pushed it from the soil, (where by gravitation it should have remained,) and directed its development upwards and laterally, forming its fruit-bud, blossom and fruit, and holding the latter suspended in the air, unaffected by rain, hail or wind, till in its maturity, its growth completed, it fell to the earth, by the attractive power of the same force which had repelled its parent tree from the soil. Had he done so, we might not now be compelled to begin anew the study of terrestrial physics, after having abandoned the learned speculations of this celebrated philosopher

Now, in the animal kingdom, we will begin with man, who, we flatter ourselves, is the highest development of animal life. As he stands erect upon his feet, if we suppose a vertical plane to be passed through his person laterally, the curved line so produced, tangential to his prominences, would be an ellipse. The revolution of that ellipse, on its longer axis, would produce an ellipsoid. Now, that ellipsoid is, during the life of the man, a magnet, with opposite poles at its head and feet, and various parts of his body are also separate magnets, but in harmony with the chief magnet. His legs are a horse shoe magnet, with the poles in the feet, and the five toes on each of his feet constitute, for each foot, four horse shoe magnets. When, from disordered health, the magnetism in either leg is no longer produced, paralysis of that limb results, and the contractile and expansive power of the muscles is no longer acted upon by the electricity of the system. The arms furnish another horse shoe magnet, and the five fingers of each hand constitute, each, four horse shoe magnets, with the poles at the extremities. The optic, nasal and auditory nerves, in each pair respectively, constitute a horse shoe magnet. The genital organs are each a separate, but very powerful magnet, and are ellipsoids in form.

In quadrupeds, the fore legs are a horse shoe magnet, as also are the hind legs. The split hoofs of the ruminants are also horse shoe magnets; so are the round hoofs of the horse, the ass, the mule and the zebra, with their poles pointing to the rear, instead of to the front. A lateral horizontal section of a quadruped through his head, neck and body, would develop an elliptical curve. The jaws of animals are separate horse shoe magnets. A serpent, which is also an ellipsoid, is a magnet, and when it is coiled, each of its coils preserves the ellipsoidal form. The same type runs through the feathered tribes, and the forms of the fishes everywhere partake, more or less, of the elementary character of the ellipsoid.

In the investigation of this subject it will be found that the attachment of animals to the earth, and their locomotion upon it, are due to magnetism, and not to gravitation. It will be observed, that in all animals, their bodies, which are their heaviest parts, are the farthest removed from the surface of the earth, which could not be the case if they were held to the earth by the attraction of their weight or gravity. As Newton's rule is that the attraction of gravitation is proportional to the mass or weight, and, as the head, neck, body and thighs are the heaviest parts of the animal, they should be nearest to the earth, which it is known. they are not.

Now, why is this type so universal—as well in planets as in whatever that has life upon them? Is it not because of magnetism, that has developed this form and its modifications? Does not the magnetism of the atmosphere control the movements of birds by its attractions and repulsions; of the sea, which is highly magnetic, those of the fishes and marine animals which inhabit it; and of both the air and the land, those of the animals who live upon the land, and of the plants which are developed in its soil? Magnetism, therefore, is an element of life, in plants and animals, and is one of the motive powers of planetary and stellar movements in the universe.

Let us now return to Moses and his book of Genesis. In the 2d chapter and 7th verse, he says: "And the Lord God formed man of the slime of the earth, and breathed into his face the breath of life; and man became a living soul." And in the 21st verse, "Then the Lord God cast a deep sleep upon Adam, and when he was fast asleep, he took one of his ribs and filled up flesh for it." And in the 22d verse, "And the Lord God built the rib which he took from Adam into a woman, and brought her to Adam." When we remember the history of Moses, his birth of Israelitish parents, in the province of Goshen, bordering on the Delta of the river Nile; the attempt of his mother to save him from the destruction decreed by Pharaoh against all the male children of the Hebrews, by placing him on the river Nile, in a water tight cradle made of papyrus, among the water plants of that stream; his discovery by Pharaoh's daughter as she was proceeding to bathe in the river near by; his delivery to his mother to be nursed and reared, till he should be old enough to be educated as the adopted son of the Princess, who had discovered him in the river; his education by the priests, who at that period, as a class, were the most learned persons in Egypt; his subsequent abandonment of the court of Pharaoh, and flight into the desert, where he passed forty years of his life; his selection as leader of his people in their flight from Egypt, and his residence among them for the last forty years of his life; we are not surprised that so learned a man, of such varied experiences, should have been chosen to conduct such a people as the Israelites out of bondage, to a land flowing with milk and honey.

In the temples of Egypt, he had doubtless seen the priests oftentimes engaged in making their idols out of the slime of the river Nile. Perhaps he himself may have assisted in their manufacture. He must have had the history of his life imparted to him, and the ooze of the river on which his cradle had rested must have been to him a familiar object. He knew the plastic character of its slime, how easily it could be made to assume any form. And he was probably acquainted with the qualities of the various materials composing it, viz: the carbonate of lime, from the bed of the river, the remains of fish and reptiles, replete with phosphates, and the vegetable matter, in almost every stage of decomposition. When, therefore, it was revealed to him by the Almighty that he had formed man out of the slime of the earth, he could readily understand that Divine power could fashion a man out of such materials, but the investing this man of flesh made of clay with life, by simply breathing into his face, was such a manifestation of power as must have confounded all his reasoning faculties.

Let us see if we can form any idea of how this vitalization of the first man was effected. Remember that this is a revelation of a physical fact, and in communicating it to mankind through the medium of Moses, the Creator did not mean to make any secret of it, but has left it to us to discover, if we can, without discrediting the act, or disbelieving the revelation. Let us suppose the first man to have been made out of the materials mentioned. He is complete in all his organisms; they are all prepared and ready to work as soon as vitality shall be imparted to them. This is done by "breathing in his face the breath of life," and "the man becomes a living soul." Now, the first inquiry is, what is the breath of life? According to Moses, light had been created, the earth had received its form, the three kingdoms, animal, vegetable and mineral, were defined, and their functions were being performed, an atmosphere existed, and we may suppose that it was constituted to fulfil all the conditions which appertain to it at the present day. Its elements were the same then as now. Light, which from the beginning had been passing through interstellar and interplanetary spaces, with its inconceivable velocity, had, on entering the denser medium of the atmosphere, produced enormous friction, by which electricity, and subsequently magnetism, had been evolved to perform the parts assigned to them in the Divine economy. When Adam, therefore, was finished in his structural condition, and the blood lay in his heart and lungs, arteries and veins, without motion, but ready for use, all that was necessary was to fill his lungs with atmospheric air, negatively electrified, and life at once became established in his system. This was done by breathing in his face the breath of life, that is to say atmospheric air, which, conducted by the nostrils and the mouth through the windpipe to the lungs, and through the eyes and ears to the brain, and meeting there the blood oppositely electrified, the conjunction of these opposite electricities produced heat, which, consuming the carbon of the blood in the oxygen gas of the atmospheric air, formed carbonic acid gas, thus purifying the blood of its carbon, imparting to it a heat of 100° of temperature, positively electrified, and expelling from the lungs, through the mouth and nostrils, the carbonic acid gas which has been thus formed. The blood, after having been thus purified, rushed into the heart, driven by the positive electricity of the lungs, and from the heart forced into the arteries, from which it was distributed to all parts of the system for its renovation and support. This arterial blood, starting from the heart with a temperature of 100° F., rolls in the arteries, producing friction and evolving electricity, supplying all the organs of the body with various materials for their renovation and nutrition, and developing magnetism, but losing more heat than it generates, so that by the time this arterial blood has passed through the capillaries, and has entered the veins to return to the heart, it has lost two degrees of temperature, and it returns to the heart as venous blood, with a temperature of 98° F. This loss of two degrees of heat in traversing the body, changing the electricity of the blood, by induction, from being positive to being negative; in the heart it becomes again positive, and rushes into the lungs to meet the negative electricity of the atmospheric air, where the same process of burning the carbon of the blood in the oxygen gas of the atmospheric air, purifying the blood, driving it back again into the heart and thence through the arteries throughout the system as before, and so on while life exists in its normal condition. This is, probably, the physical life of man, as described in the 2d chapter and 7th verse of the book of Genesis; and we find that electricity, heat, and magnetism, are essential elements of it, and that without them it cannot exist

Dr. Ure, in his celebrated experiment of conveying currents of electricity along the spinal nerves of the recently executed malefactor, Clydesdale, while the body was still warm, though life was extinct, produced a horrible caricature of the operations of life, by calling into violent contractions the muscles of the face. All the expressions of rage, hatred, despair and horror were depicted upon the features, producing so revolting a scene that many spectators fainted at the sight. In like manner muscular contractions and expansions of the limbs, imitating the movements of actual life, were exhibited, to the astonishment of beholders.

The ingenious physicist, Ritter, of Munich, in Bavaria, celebrated for his experiments in galvanism, has, through them, among other things, established the fact, that a constant development of electricity accompanies all the phenomena of life. Now, as magnetism is developed by currents of electricity, it follows, that in moving the legs of animals the expansion and contraction of their muscles produce friction and evolve an electricity opposed to that which has set them in motion, and, at the same time, the conjunction of these opposite electricities also develops magnetism, which at once is acted upon by the superior magnetism of the earth, and hence you have a leg lifted from the earth and another placed upon it, in locomotion, by the force of magnetism, and this is repeated and continued at the will of the animal.

The celebrated naturalist, Prof. Louis Agassiz, in his lectures on Embryology, stated, that the beginning of animal life was in an egg. Let us see if we can comprehend its transmutation into life. The sexes are oppositely electrified. In the human race the females, from the positive and persistent character of their demands, may be termed positively electrified. The males, from their habit of negation or denial of the wants of the females, which is of too common occurrence, may be termed negatively electrified. These opposing conditions create sexual attraction; when a conjunction of these opposite electricities occurs in the act of coition, a certain degree of heat is developed, and magnetism is also evolved—the egg disengaged from the ovarium is magnetized and positively electrified, and through the Fallopian tubes, enlarged by the heat of the coition, is carried into the uterus, prepared to receive it. Thus, vitalized by the electricity and magnetism that have been imparted to it, its own heat, and that of the uterus, in which it is deposited, continue to preserve the life which has thus been called into being. Such, also, is the commencement of animal magnetism.

Du Bois Reymond states "that the electrical current manifests itself in different directions, in the limbs of different animals, and with greater intensity in some animals than in others. The electro-motive forces thus operating in the muscles depend upon the opposite electrical [? magnetic] conditions existing between their longitudinal and transverse sections." So, also, with respect to the nervous system, he states that the nerves are subject, in their sectional arrangements, to the same law as the muscles. This must be understood, how ever, with reference only to the exercise of their inherent electro-motive forces. In transmitting the muscular current the nerves perform the part of inactive conductors. It is not in the whole, or a large part of a muscle, that an electrical current can alone be shown to exist, but that every particle, the merest shred or fragment, even what may be considered microscopic, is equally obedient to electrical influence. * * * * Every movement, look or gesture, every sensation of pain or pleasure, every emotion however transient, and perhaps every thought unexpressed, or word uttered, is most assuredly accompanied by the disturbance of electro-motive forces. These, however, are so much more feeble than any with which we have hitherto become acquainted, that in the healthiest and most active, during a week, or perhaps a month, their cumulative effects may not be equal to those evolved by one smart blow of the hand upon a table."

Much speculation has been evoked and various experiments at different times instituted, to discover and explain the cause of the uniform normal heat of the body of a healthy adult person, but heretofore with unsatisfactory results. Now, it seems to me that the explanation is not a difficult one. It will be admitted that the relative capacity of the lungs to furnish atmospheric air to oxidate the blood, and of the heart to supply the proper quantity of blood to be so oxidated in the lungs, is constant in a healthy adult. When, therefore, the lungs are filled to their greatest capacity, with blood and atmospheric air in diffusion through it, the meeting of the negative electricity of the air with the positive electricity of the blood in the lungs, develops heat and magnetism, and the oxidated blood becomes positively electrified; the carbon of the blood unites with a portion of the oxygen of the air in the lungs, and becomes carbonic acid gas, also positively electrified. This change also develops heat and magnetism, having been produced by the meeting of opposite electricities; a portion of the water of the blood, separated from it during these changes, is taken up by the carbonic acid gas; and the carbonic acid gas and the oxidated blood, both being positively electrified, repel each other—the blood back to the heart, to be thence distributed by the arteries through the system, while the carbonic acid gas, and the watery vapor it contains, are expired from the lungs through the mouth and nostrils into the atmosphere. This repulsion of the carbonic acid gas and watery vapor from the lungs is obvious to every one. For who is there that can hold his breath even for a single moment? A greater power than man's will forces them from the lungs, and that is the repellent power of positive electricity. The oxidated blood is driven into the heart by this same repellent force.

It is the electrical action, therefore, in the lungs of the atmospheric air and the blood intermingled in constant relative quantities, that produces the uniform temperature, in all latitudes, of 98° Fahrenheit in a healthy adult person.

Electricity is the cause of the fluidity of the blood in the veins and arteries. Venous blood taken from the veins, and left to itself becomes solid, and separates into two distinct parts; the serum, or watery, being over and upon the clot or coagulum. The serum is chiefly water, holding albumen in solution and the salts of the blood. The clot contains fibrin coloring matter, a little serum and a small quantity of salts. Prick a finger with a needle, a small drop of blood exudes. It is negatively electrified; on being exposed to the air its negative electricity instantly unites with the positively electrified air in contact with the warm surface of the finger, heat is produced by their conjunction, the watery part of the serum is evaporated by the heat and the distributing electricities; and the clot remains to cover the puncture made by the needle, and to protect the blood in the vein from further injury by the action of the air upon it. How many lives have been saved after unconsciousness, from the loss of blood in wounds, has seized upon the sufferer, by the escape of the serum of the blood through evaporation from electricity, and the deposit of the clot upon the lips of the wounds, closing them and preventing the further flow of the blood through them, and thus allowing nature to gather up its remaining strength, and to restore the patient. How thankful we should be to the Creator for this simple, wise and benevolent provision for our safety in the occurrence of blood-letting injuries!

An eminent surgeon of my acquaintance has informed me, that, in cases of death produced by lightning, the blood remains fluid in the veins for several days afterwards; whereas, in cases of death from disease, the blood coagulates soon afterwards. He has known a case in which the blood remained fluid in the veins four days and several hours subsequent to the death of the man by lightning. This goes to show that in the absence of electricity from the blood, its flow in the arteries and veins becomes retarded, and its coagulation, or even thickening, would suddenly terminate the life of an animal in which it had occurred. This, no doubt, is the cause of paralysis and appoplexy. The treatment in such cases, therefore, should be the introduction of the opposite electricity in the veins and arteries to restore the electrical equilibrium and consequent fluidity of the blood.

I have somewhere met with the following anecdote of the late Emperor Nicholas I, of Russia, which, as it is pertinent to the present discussion, may be introduced here. It is as follows:

Some years since a very distinguished French actress, having an engagement at the Imperial Theatre at St. Petersburgh, arrived there at the beginning of winter. Soon after her arrival, in company with a gentleman of her party, she proceeded to the grounds of the Winter Palace for walking exercise. Winter had arrived and the ground was covered with snow, some of which had recently fallen. The air was calm and the weather very cold.

In the course of their walk, their attention was attracted by the appearance of a gentleman of very distinguished mien, who was also walking. He was very tall and remarkably handsome, and was approaching them rapidly; very much impressed by his appearance and manner, they were regarding him very fixedly, when as he came near to them they saw him take off from his hand a glove, and stooping low he grasped a handful of the light and newly fallen snow. This strange movement so fully occupied their attention, that they were almost unaware of his having reached them, when, stopping before the lady, he very abruptly clapped his hand filled with snow upon her nose, and began to rub it vigorously, at the same time saying to her in French; "Madame, your nose is frozen!" Her attendant, astounded by what at first he thought was intended as a great indignity to the lady, was about to resent it, when he heard the explanation which accompanied it. The Emperor Nicholas, for it was he, began to rub briskly the nose and face of the lady with his hand filled with snow, to restore, by friction, the proper circulation of the blood, and thus prevent the great injury to the lady's face which the loss of her nose would occasion. He spoke encouragingly to her, and calling an attendant he sent for his surgeon, and after the circulation of the blood in her face was re-established, she was returned to her apartments, where she received every attention, by the Emperor's orders, and in a little while she was entirely restored. Now, why did the Emperor rub her nose and face with snow; and why did he take off his glove from his hand to perform that office?

It has been long known, that frozen limbs can be restored to their normal condition of healthy vigour by the application of snow or pounded ice to the part affected, when quickly rubbed with the human hand; but it is not so well known why such an effect is thus produced. Let us essay an explanation of it. When a limb or member is frozen, the circulation of the blood in it ceases, and the life of the limb or member is suspended; and unless its healthy action is speedily restored, the part affected loses its vitality, gangrene sets in, and amputation becomes necessary. The animal electricity that it contained has disappeared. Now, the human hand has one kind of electricity; snow or ice has the opposite kind of electricity. When these opposing electricities are brought together in contact by friction, as they were in this instance, heat and magnetism were evolved, which heat warmed and expanded the frozen nose, and associated with the magnetism that had been developed, excited an electrical current in the coagulated blood in the veins of the nose and face, which then began to flow in its natural course. When this friction is thus continued for a sufficient time, the health of the limb or member is restored. Now if heat from combustion had been applied in this case, instead of heat from electricity evolved by friction, as above described, it would have resulted in the mortification and loss of the lady's nose.

It has been abundantly shown, by experiments made by distinguished scientists, that, under the influence of weak currents of electricity, salts can be resolved into their component elements. In this way a compound can be separated into its constituent acid and base. It has also been shown, by Becquerel, that if an acid and alkaline solution be so placed that their union is effected through the parietes of an animal membrane, or, indeed, of any other porous diaphragm, a current of electricity is evolved. This has been found to be true with all acids and soluble bases. Now, Dr. Golding Bird asserts, that "with the exception of the stomach and cœcum, the whole extent of the mucous membrane, is bathed with an alkaline mucous fluid, and the external covering of the body is as constantly exhaling an acid fluid, except in the axillary and pubic regions. The mass of the animal frame is thus placed between two great envelopes, the one alkaline and the other acid, meeting only at the mouth, nostrils and anus. Donné, has shown that this arrangement is quite competent to the evolution of electricity.

"The blood in a healthy state, exerts a well marked alkaline action on test paper—but a piece of muscular flesh containing a large proportion of alkaline blood, when it is cut into small pieces and digested in water, the infusion thus obtained is actually acid to litmus paper. This curious circumstance is explained by the fact announced by Liebig, that, although the blood in the vessels of the muscle is alkaline from the tribasic phosphate of soda, yet the proper fluids or secretions of the tissues exterior to the capillaries are acid from the presence of free phosphoric and lactic acids. Thus in every mass of muscle, we have myriads of electric currents, arising from the mutual reaction of an acid fluid exterior to the vessels or their alkaline contents. It is thus very remarkable, that a muscle should be an electrogenic apparatus, and that we should have two sources of the electricity of the muscles—the effects of metamorphoses of effete fibres on the one hand, and on the other the mutual reaction of two fluids in different chemical conditions. The agency of a muscle ingenerating electricity can no longer be denied.

"In the course of twenty-four hours a considerable proportion of watery vapour exhales from the surface of the body. This has been differently estimated, and is liable to great variations—but from 30 to 48 ounces of water may thus be got rid of from the system. The evaporation of this amount of fluid is sufficient to disturb the electric equilibrium of the body, and to evolve electricity of much higher tension than that set free by chemical action. This evaporation may probably account for the traces of free electricity generally to be detected in the body, by merely insulating a person and placing him in contact with a condensing electrometer. Pfaff and Ahrens generally found the electricity of the body thus examined to be positive, especially when the circulation had been excited by partaking of alcoholic stimulants. Hemmer, another observer, found that in 2422 experiments on himself, his body was positively electric in 1252, negative in 771, and neutral in 399. The causes of the variations in the character of the electric conditions of the body, admit of ready explanations in the varying composition of the perspired fluid. For if it contains, as it generally does, some free acid, it, by its evaporation, would leave the body positively electric; whilst if it merely contains neutral salt, it would induce an opposite condition. The accuracy of these statements can be easily verified by means of the electrometer."

"It is an established fact that, independently of combustion, chemical action or evaporation, the mere contact of heterogeneous organic matters is competent to disturb electric equilibrium."

"Whatever may be the influence of electricity as an agent in exciting the function of digestion, it is now pretty distinctly made out that the function of digestion in the stomach is an action allied to simple solution, of which water—a proper temperature, [always associated with electricity]—and a free acid, the hydrochloric, phosphoric, or both, are the active agents. We possess sufficient evidence to induce us to regard a current of electricity as the means by which the saline constituents of the food are decomposed, and their constituent acids, the real agents in digestion, set free in the stomach, the soda of the decomposed salts being conveyed to the liver to aid the metamorphosis and depuration of the portal blood, and cause the separation of matter rich in carbon in the form of a saline combination in the bile. It also appears, from various experiments, that in all cases the secreted matters are always in an opposite electric condition from that of the blood from which they were generated."

Chemical action is merely a synonym for electrical action, hence in all the functions of the animal body from its birth till its dissolution, we may observe the influence of electrical currents, the development of magnetism by the conjunction of them, oppositely electrified, and the production of heat. In the first inspiration of atmospheric air into the lungs where it encounters the blood oppositely electrified, heat and magnetism are evolved, and the purified blood has one electricity which repels itself into the heart, and thence by the arteries through the system. When it reaches the capillaries it has lost more than two degrees of its temperature, and being forced through the capillaries into the veins as well by the repulsion of the electricity of the arterial blood, as attracted by the opposite electricity of the veins and the blood they contain, the temperature is increased till it reaches 98° of Fahrenheit, which it carries with it to the heart.

Muscular exercise actively employed by the contraction and expansion of the muscles, and by their friction among themselves, develops large quantities of electricity, which requires a corresponding quantity of the opposite electricity of the air to neutralize it, hence the inspiration of atmospheric air into the lungs becomes more rapid in proportion to the activity of the exercise, great heat is developed in the body by the conjunction of these opposite electricities, which expanding all the tissues of the body, liberates the water contained in them and in the viscera by exos mosis, which then exudes through the pores of the skin, as perspiration, carrying off the surplus electricity that has been produced by the violence of the exercise, and relieving the body from the further inconvenience of its increased heat. This perspiration is acid in some parts of the body and alkaline in other parts, and furnishes the most immediate means of getting rid of the excessive free currents of electricity of the body at all times.

During an attack of fever, while the patient is suffering from the great interior heat of his body from disturbed electrical action, why does he continually ask for cold water? It is because the cold water, oppositely electrified to the overheated organs and viscera of his body, is demanded by the instinct of his nature, which requires it, so that the increased heat developed by the conjunction of these opposite electricities may still more expand the tissues and viscera and liberate the water therefrom which, mixed with the water drank, would carry off in perspiration the excess of electricity and restore the body to its normal condition. For this reason, cold water in large quantities should always be prescribed in cases of fever, to carry off the surplus electricity, by the perspiration it induces, as well as to supply the material for the very perspiration that it is intended it should produce. Warm saline or acid baths, by expanding the pores of the skin, and thus promoting perspiration, are natural remedies in cases of fever or of violent inflammation. Perspiration, therefore, alkaline or acid, is the remedy for excessive electrization—and just as the perspiration is either alkaline or acid, in those places of the body where in its natural state it should be the reverse, ought the physician to be able to diagnose the causes of this abnormal condition, and to restore the electrical equilibrium in the system.

The sexes are oppositely electrified—hence their mutual attraction for each other. Now give them the same electricities, and mutual repulsion immediately results. Let us ponder awhile on this subject. Every one must have observed in the press of this country, almost daily, and in every part of it, accounts of the most outrageous, cruel, and in some cases of diabolical attacks of men upon women, and occasionally of women upon men, generally when they bore toward each other the relation of husband and wife. When they have been first acquainted with each other, their electricities being opposite, they were mutually attracted to each other, their acquaintance grew into esteem, and ripened into affection and love, and they became man and wife. The animal system develops electricity, magnetism and heat in its functional actions—the kind of electricity and magnetism are dependent upon the habits of life, the diet, the occupation and association of the individual. When these are similar similar electric and magnetic conditions of the body will result. It has been shown that the negative or masculine electricity of the man is reversed, and becomes positive like that of the woman under the excitement of alcoholic stimulants—in other words, for the time being, the man becomes a woman, and is converted into the only thing which the British Parliament, in all its great potentiality, could not do, viz: make a man a woman, or a woman a man. This, alcoholic stimulants have always done, and are now doing every day. When this change in the condition of his electricity has occurred, his attributes become feminine; he is irritable, irrational, excitable by trivialities, and when opposed in his opinions or conduct, becomes violent and outrageous, and if, in this mood, he meets his wife, whose normal condition of electricity is like his present condition, positive, they repel each other, become mutually abusive, engage in conflict and deadly strife, and the newspaper of the next day announces the verdict of the coroner's jury on the case How many such incidents are occurring daily in almost every part of our extended country; and who would expect to find the discovery of the moving cause of all these terrible crimes in the perspiration of the criminal? and yet science has shown that the metamorphosis of a man into a woman by changing the negative condition of his electricity into the positive electricity of the woman, with all its attributes, is disclosed by the character of his perspiration, superinduced by the use of alcoholic stimulants! It is a very curious thing to note, that among the Persians, one of the most ancient of peoples, the ordinary salutation on the meeting of friends, is, not as among the English, "How do you do?" as if your life was one of incessant labor, or as among the French, "Comment vous portez-vous?" "How do you carry yourself?" as if it was a great exertion to move at all—but "How do you perspire?" In the lapse of ages, a vast deal of knowledge useful to a people, is necessarily acquired by their experience, personal as well as national. In the hot and arid climate of Persia, the people suffer, and have always suffered, greatly from fevers, eruptive diseases of the skin, as well as from those of a dysenteric and choleraic character. Their experience has taught them, in their diseases, that the first relief from suffering that they felt, was in the return of their perspiration to their skin, and as long as that perspiration could be maintained, just so long was their relief continued—hence they came to regard it as synonymous with a state of good health, and the salutation among friends on meeting was introduced and became common among the people.

Let no woman, hereafter, delude herself with the idea that she can reform a man addicted to the use of alcoholic stimulants by marriage. Should she attempt it, she will fall a victim to the delusion, as many of her sex have done before her, as she will find that her will is controlled by her normal positive electricity, which is of the same character as that of the man, her husband, and that, in spite of herself, the two will be mutually repellent, and their association as man and wife will be unhappy in the extreme.

Observe a drunken man with a male companion who is sober; their electricities are opposite; how loving the drunken man is to his friend; he caresses him; locks his arm in that of his companion; hugs him; in France he would kiss him; prattle to him with the simplicity of a child; talks nonsense with the incoherence of delirium; and is as good humored and amiable as possible. His wife appears on the scene; his manner changes instantly; she tells him he is wanted at home, and asks him to accompany her there; he replies, "you go to grass, don't you see I am with George," naming his companion. The wife urges him to go home, and not expose himself in the public streets in his condition. He is exasperated; their repellent electricities are in action; they become angry; violence probably ensues, and the police interfere. Let no woman ever venture to remonstrate with a drunken man; her own electrical condition forbids it; such remonstrance irritates the man, develops his anger, and leads to violence; and when it is remembered that women are particularly the objects of brutal attack by drunken men, as is made manifest by the publication in the daily press of the country, of crimes that have been committed, it is obvious that their safety will be promoted by their silence.

The remarkable variations in his own electrical condition, reported by the observer, Hemmer, as deduced from his experiments upon his own body, go to show that every incident in human life might be traced to its electrical condition; all the passions are excited by it, and are subdued by its reversal; all the emotions are necessary consequences of it, and it is not probably going too far to say that the intellectuality of man is largely due to his electricity and magnetism.

We have thus shown that from the impregnation of the ovum of the warm-blooded animal, through its whole existence, electricity, magnetism, and heat, are the essential elements of its vitality; and that starting from the first man, Adam, it was not until the Creator had "breathed into his face the breath of life," or, as we interpret it, had brought together the atmospheric air and the blood in his lungs, oppositely electrified, by breathing that atmospheric air into his face, t rough his mouth, nostrils, and eyes, and thus bringing it into contact with the oppositely electrified blood, that life in Adam was established, and the law of life made universal for all his descendants.

It is curious to observe the marvelous provisions made by the Creator to relieve the human animal from the excess of electrical action in his system from whatever cause. The brain being the most important of the organs, and contained in a bony structure called the cranium, or skull, composed of several parts united by serrated edges, and subject to a certain degree of mobility at those edges, to protect the skull from fracture by trivial, accidental blows, or pressure, is the first organ to be relieved from increased heat in the blood which circulates there. Perspiration first breaks out on the forehead, near the temples; then at the uppermost suture, or serrated edge, on the top of the skull; then along the temples; then behind the ears, to relieve the cerebellum and the organs of hearing; then above and below the eyes, for the relief of those organs; then along the nose and corners of the mouth; then under the jaws, to relieve the glands of the mouth and throat; the thorax, or chest, where the greatest activity of the circulation of the blood occurs, is relieved by the perspiration in the armpits, under the shoulders; while the abdominal region is protected by its exudation in the loins and groins, and the pelvis and hips have their guardian in the pubic region; the upper leg in the angle behind the knee, when it is bent; the lower leg and foot find their security in the perspiration that exudes between the toes, as the lower arm and hand are protected by it, as it escapes between the fingers and in the palm of the hand—all these salutary provisions are independent of the will of the individual, and are so many safety valves for his preservation from injury, in too many cases, from his own imprudence and folly.

It is to the female of every species that the Creator has confided the care and perservation of the young animal, as well as the continuance of the species to which she may belong. We all know how powerful is the emotion of maternal instinct; it needs no illustration.

Among all animals but man the season of reproduction is dependent upon climatic influences—upon the temperature of the season, when the young animal is to be ushered into life, and on the products of the earth necessary for the mother during the period of its dependence upon her for sustenance as well as for its own support afterwards.

We will illustrate by a common example. We will suppose

that the season for reproduction with the domestic cow has arrived; she is at pasture, and unconscious of the change in her condition which is about to happen. Suddenly, there begins to be given out from her body a strong effluvium—it surrounds her and accompanies her in every movement. It fills the atmosphere near her—wafted by the wind it is carried to a great distance. A mile or more to the leeward of the cow, a bull is feeding among a hundred cows, in the pasture field; grazing quietly he is observed to turn his head towards the direction from which the wind is coming. It marks the first approach of the effluvium; he turns quickly around towards the wind, raises his head high above his body and draws a long inspiration of air. He recognizes the fragrance. It is to him an invitation. He sets out in a rapid walk in the direction from which the wind is coming; then he quickens his pace into a fast trot, and, as the welcome perfume increases in strength, he breaks into a gallop, and then into a full run. A fence, a barrier, intervenes; raising himself on his hind heels he throws his forehand on the fence and breaks it to the ground. Renewing his speed he arrives in the field in which the cow is quietly grazing—among a thousand cows. He follows the fragrance directly to the object of his visit. Now, what does this haste mean? Why does he leave his own pasture, a mile or more away, to rush with such speed to other fields? Because a new life is to be developed, and the indispensable elements of it are heat, electricity and magnetism. The exercise of his muscles in running has produced friction, friction has developed electricity, positive, which demands negative electricity from increased inspiration of the atmosphere. His imagination has been excited by the pungency of the grateful aroma he has breathed. He arrives at the cow, draws a long inspiration, licks her on the neck with his rough tongue, and upon her loins, and makes an effort, as Jupiter is said to have done to Europa, after crossing the Bosphorus. The cow recedes from him, and he is disappointed—she is not ready. Again and again he proffers his devotion—still rejected. The cow, in the meantime, recedes from him a few paces, and begins again to graze. Every moment, however, her maturity of passion is approaching, the circulation of her blood increases, stimulated by his proximity and the odour given out from his body. Heat and electricity in her body are developed by a quickened circulation, and when the instinct of her nature has been fully aroused she communicates to him, in a mysterious way, her readiness to receive, in the language of the Latin poet, "*taurum ruentem in Venerem*," the elements of life are there, electricity, magnetism and heat, and at the end of the period of gestation, a new life is added to the herd.

Among birds and poultry, the requisites for reproduction are similar. In the poultry yard observe the gallant cock. Scratching on the ground he finds a grain of corn, or perchance an insect; he gives a chuckle and one of his hens approaches to receive it. She picks it up, and comprehending the generous motive of the gallant bird, she starts off in a run to enjoy the gift. The cock pursues, and after a sharp and quick race, in which friction, electricity, heat and magnetism are developed in each of them, she suddenly stops, an embrace follows, and an egg is impregnated, which in due time is hatched into a chicken.

Sometimes, the cock pretending to have found some choice morsel when in fact he has not, calls a hen, who on approaching him discovers the cheat and starts from him on a run, to be pursued by him as before, and with precisely a similar result to the last mentioned. So that to be a gay deceiver of the female is not confined to base man.

In the reproduction of all the varieties of animal life, from the enormous whale to the firefly, which in the language of Tom Moore, "lights her mate to her cell," and from it to the tiniest insect, the like conduct prevails, viz.: the exercise of the muscles producing friction, and evolving electricity, magnetism and heat, to vitalize the ovum in its impregnation.

The whale requires three-quarters of an hour to be passed in sportive dalliance around his mate, before a sufficient degree of electricity, magnetism and heat can be attained to impregnate the ovum of the female.

I have been credibly informed by a very intelligent man, who was for many years engaged in the whale fishery in the Southern Pacific ocean and Australian seas, that while cruising for whales off the coast of Australia the boats of his ship pursued and captured a large sperm whale that made 90 barrels of oil. That when first struck with the harpoon he went down with great velocity, carrying with him an immense length of line, and that before he arose again to the surface "to blow" *one hour and twenty-three minutes by the ship's chronometer* had elapsed, which fact proves that it is not necessary for a whale to come to the surface of the water at short intervals of time to breathe, as naturalists suppose, as from the lapse of time mentioned while he was under the water he evidently had supplied himself with atmospheric air for breathing purposes from the water, as it was impossible that any pair of lungs could have inhaled and retained sufficient air before he went down to sustain him for so long a time under water. The true explanation probably is, that the whale came to the surface to blow off, with his carbonic acid gas and watery vapour from his lungs, the surplus electricity that had been evolved in his system by the immense muscular action he had displayed in his descent from, and subsequent ascent to the surface, as by no other method could he have gotten rid of it.

Among terrestrial animals nothing is more common during the heats of summer, when so much electricity is evolved within them by their inspiration of air, the circulation of their blood, their digestion, secretions and muscular action, than to see them in herds standing in water up to or above their knees to relieve themselves of their surplus electricity by the conducting power of the water and thus to cool their bodies whose heat must ascend into the air, and could not be conducted to the earth while their electricity could, by the water in which they stood, be rapidly conducted from their bodies to the earth.

Such is likewise the cause of the habit of wallowing in muddy water of all the pachydermata, from the mammoth through the elephant, rhinoceros, down to the common pig.

All fatty or oleaginous substances being anti-frictional, as is illustrated in every day life in the axles of our vehicles and in machinery having any rotating associations, prevent the evolution of electricity, and consequently of heat. Hence some extraordinary facts appear in the animal economy. It is known that the whale, one of the varieties of the cetacea, nurses its young from its teats, which are external on its body. It is therefore classed, by naturalists, with the mammalia, to which the human species belongs. The whale inspires atmospheric air, when floating on the surface of the water, and also abstracts it from the water itself when swimming beneath its surface. The whales are warm blooded, and the conjunction of the negative electricity of the atmospheric air they have inspired, with the positive electricity of their blood, produces heat. This heat and the accompanying electricity, which is derived from the friction of their blood in circulation, and of their muscles in exercise while in motion, would all be rapidly conducted from their bodies by the water of a lower temperature, in which it moves and lives, but for the great thickness of the blubber or fat which encompasses them respectively, and the immense quantity of oil contained in their skulls, that are non-conductors of electricity, and serve to insulate it as it is evolved. How then, in the rapid passage of a whale through the water, is the enormous quantity of electricity evolved by the friction of its organs, muscles and blood, in their respective motions, to be got rid of since it cannot escape from its body on account of the non-conducting power of the robe of blubber which encloses it? The whale, in breathing, takes in a large quantity of water containlng atmospheric air, which air, having one electricity, is received into its respiratory system, where it meets with the blood oppositely electrified. This blood it oxygenates, and by the positive electricity of its lungs and heart, this blood, similarly electrified, is driven through the arteries, to carry to every organ of its body its renovating and vitalizing material. Changing the character of its electricity by induction as it passes into the veins, through the capillaries, it is taken back to the heart and thence to the lungs by the attraction of the positive electricity of those organs, to maintain the life of the animal, and this process is continued during its existence. Now the air which the whale has inspired, whether from the atmosphere directly, or by abstraction from the water in which he lives, after it has been used to oxidate his blood, is to be gotten rid of. But how? This air being warm carbonic acid gas, and associated with watery vapour produced by the heat of opposite electricities in converting the carbon of the blood into carbonic acid gas during the act of breathing, is positively electrified, and is repelled from the lungs by their positive electricity, into the atmosphere negatively electrified, through its blow holes or spiracles, and thus the act of breathing among animals is nothing more or less than the action of electricities in their opposite condition of attraction and repulsion, when associated with inspired and expired atmospheric air.

Professor Matteucci has incontestably proved, "that currents of electricity are always circulating in the animal frame, and are not limited merely to cold blooded reptiles, but are common to fishes, birds and mammalia." He has shown that a "current of positive electricity is always circulating from the interior to the exterior of a muscle, and that muscular contractions are developed in the animal machine by a fluid which is conducted from the brain to the muscles."

The contraction of a muscle is produced by an electric current of one kind. The extension of it is occasioned by another current of opposite electricity. These alternate forces, applied to the muscles of an animal, keep them in healthy exercise, and occasion all their movements, whether voluntary as directed by the will, or involuntary as independent of it. When a person, therefore, is immersed in water, particularly in sea water, he is apt to be drowned; for the positive electricity which flows from the interior to the exterior of his muscles, extending them, is carried off rapidly by the negative electricity of the water in which he is immersed, leaving the negative electricity flowing from the brain to the muscles, to contract them in cramps, which he is not able to overcome, as he has lost the power to extend his limbs by the escape of his positive electricity into the water. This is the cause of the frequent drowning of persons; even the best swimmers are sometimes drowned from this cause. The Creator has provided a remedy against this loss of positive electricity in aquatic birds; covered with down and outside feathers, they secrete a certain oily matter with which these birds, puncturing with their bills the vesicles containing it on the surface of their bodies, and filling their bills with it, anoint their feathers, rendering them impenetrable by the water in which they swim, and thus they retain not only their electricities but also the necessary temperature of their bodies which the union of these electricities in their bodies develops. The women of the South Sea Islands, in the Pacific Ocean, having taken the hint from these birds, without comprehending its reason, when they go to swim anoint their bodies with palm or cocoanut oil, and boldly plunge into the sea, swimming a mile beyond the breakers which surround their island homes, and taking with them a piece of board, sufficient to bear their weight, on which they mount, and then standing on the board on one foot, balancing their bodies upon it, they allow the immense rollers from the ocean to bear them with great rapidity to the breakers, where thrown from their boards by the violence of their motion they swim to the shore, repeating in this manner their sport for hours, defying cramps, preserving their electricities, retaining the natural heat of their bodies, and revelling in the joyous excitement of their dangerous sports. This practice of the South Sea Islanders, it is said, has been recently imitated by the English Captain Webb, in his successful attempt to swim across the Straits of Dover, he having anointed his person before starting with the oil of porpoises, which enabled him to retain his electricity and heat in his body, and thus to accomplish his feat. Now, in cases of shipwreck, it is obvious that when people are thrown into the water, no mere floating apparatus, called "Life Preservers" are of any value to prevent the escape of the electricity and heat of the floating person; but that he is liable to be drowned in a very few minutes by the escape of those elements of life from his body, notwithstanding he may continue to float for hours afterwards. The Esquimaux and other Arctic tribes of people delight to eat oils, blubber, and other fatty substances, having been taught by their instinct that this fatty diet serves to retain within them the heat of their bodies—but how? All fatty substances are anti-frictional, and non-productive of electricity. The viscera and tissues of these fat eating people become invested with fat, retarding the evolution of electricity in their system, and by thus diminishing their interior heat, preventing the secretion of excessive perspiration, by which their electricity would be carried off from their bodies, and the consequent reduction of their temperature.

The people along the shores of the Mediterranean sea, in the south of France, Spain and Portugal, delight also in oily foods, as a preventive of the excessive secretion of perspiration, without however understanding the rationale of their diet.

The first Napoleon, in a conversation with Corvisart, his chief physician, said, that "he had no faith in the art of medicine; but that he placed a high value on surgery. Anatomy had developed a knowledge of the human organization, and post mortem dissections had displayed the effects of disease, or of injuries to various parts of the human system, by which the surgeon could profit, but that no such valuable aid was offered to the physician, who had to grope his way as best he could, in his attempts to discover the cause and the seat of the disease, and then to adopt an experimental treatment to remove it."

"But," said Corvisart, "Does your Majesty never take medicine?" "No," said Napoleon; "When I am disordered, I abstain from food, mount my horse, and ride rapidly sixty miles—on my return I bathe, sleep soundly, and the next day I am well." The rationale of this treatment is as follows, viz: The active exercise on horseback produced friction in many of his muscles, which friction evolved positive electricity; this required renewed inspiration of atmospheric air, negatively electrified, to restore the electrical equilibrium; the union of these electricities developed heat and magnetism, which conducted to the stomach and intestines served to digest the food previously taken, and which, having remained undigested, had occasioned his disorder. If any excess of electricity remained in his system after his return to the palace, the warm bath conducted it from him, and soothed him to sleep.

Solomon, the wisest of men, has left, as one of his legacies to mankind, the maxim, "spare the rod and spoil the child." Now let us examine this. When children were misbehaved, were destructive in their inclinations and conduct, rebellious to authority, and were otherwise troublesome to parents or others having the charge of them, Solomon, being a keen observer of effects, recommended personal chastisement with the rod, and naturally attributed their better deportment after the punishment, to the fear of the child of its repetition, and perhaps with greater severity. This was possibly a natural conclusion on his part, at the age in which he lived, and may be so considered even at the present time, but there is another explanation, more philosophical and more scientific. It is as follows, viz: When people are in good health, they are usually cheerful, in good humour with themselves, and amiable to those around them; they do not think of or attempt to perpetrate mischief to others, their electricities are in equilibrium, and they deport themselves properly. Now let one or other of their electricities be in excess, immediately their dispositions become changed; no longer amiable, they see every thing and person through a disturbed medium; they become sullen, cross, crabbed, quarrelsome and disagreeable; the least disappointment ruffles them, and they proceed to behave ill. Now with children, when the rod is applied vigorously to their persons, the friction produced by the blows evolves electricity of the kind necessary to restore the healthy electric equilibrium of their bodies. When that is re-established there is an end of the trouble; they become amiable and gentle. This salutary method of correcting "*les enfans terribles*," has greatly fallen into disuse in our times, from the overweening maternal instinct of mammas, which is horrified by the cries of the suffering little ones, and hence they decry against it.

This punishment is also well adapted to the adult human animal, if we are to believe a statement recently made in some of the London newspapers. It seems that the British Parliament, within a few years past, had re-established corporeal punishment with the cat-o'-nine-tails at a whipping post for a certain class of criminals, whose crimes had become alarmingly numerous. Since the re-introduction of the whipping post and its accompanying punishment, these crimes have almost ceased to exist Let other people profit by the example.

It is remarkable that three such eminent men as Solomon, Nicholas I, of Russia, and Napoleon Bonaparte, should each use in a different way the powers of electricity successfully, and yet be ignorant of the powers they were developing. Solomon by his rod correcting the wilful caprices of childhood, Nicholas I, removing the effects of frost bites, and Napoleon restoring himself to health, each by the evolution of electricity.

Let us turn now to the fourth class of vertebrate animals, which as a general rule live in the water, and prominent in this class are fishes. "A fish breathes by means of its gills, extracting the air from the water in which it lives, and rejecting the water, which carries off whatever positive electricity that may have been evolved by its muscles in its motions." This leaves the fish in a condition of negative electricity, like that of the water in which it lives, and having but one electricity, it is cold blooded—warm blooded animals having their blood warmed by the union or conjunction of opposite electricities. "Fish are nearly insensible to pain, from the same cause," as all pain in animals results from a disturbance of the electrical equilibrium of their bodies. "The temperature of fish is only 2° warmer than that of the water in which they live. They have small brains in comparison to the size of their bodies—considerably smaller in proportion than they are in birds or mammalia." This accounts for their insensibility to pain, "but the nerves communicating with the brain, are as large in fish proportionately as in either birds or mammalia. The senses of sight and hearing are well developed in fish, as are also those of smell and taste, particularly that of smell, which chiefly guides them to their food. This sense is very keen, more so than in many other animals, and thus it is that strong smelling baits are so successful in fishing."

Fish are remarkably fecund. There is nothing in the animal world that can be compared with them, unless it be some species of insects. The codfish yields its eggs in millions, from a sturgeon have been taken seven millions of eggs, flounder produces 1,200,000, the sole 1,000,000, mackerel 500,000, and so on. These eggs, if they be not vivified by the milt of the male fish, just rot away in the sea, and never come to life at all, and are of no value except perhaps as food to some minor animals of the deep.

It is now well known, that the impregnation of fish eggs is a purely external act to their bodies, fish having no organs of generation. It is this wonderfully exceptional principle in the life of fish, that has given rise to the art of pisciculture, *i. e.* the artificial impregnation of the eggs of fish, forcibly exuded from their bodies, which are brought into contact with the milt of the male fish independent altogether of the animal.

The principle of fish life which brings the male and female fish together at the period of spawning is unknown. Some naturalists have supposed that the fish do not gather into shoals till they are about to perform the grandest action of their nature, and that till then each animal lives a separate and individual life; but this does not suggest the attraction which brings them into this association.

I will venture upon an explanation. Their instinct teaches them that their eggs, when ready to be discharged from their bodies, must be deposited in warmer water than that in which they habitually swim. Having but one electricity, the negative, which is the same as that in which they live, no vivification of their eggs could take place if duly commingled with the milt of the male fish in mid ocean, but attracted by the warmer water of rivers at their sources, or in lochs or bays sheltered from the waves of the sea, where in their shallows vegetable food is always growing at the bottom for the support of the young fry, when they shall be hatched, they hasten in immense shoals for mutual protection from their enemies, to these lying-in places, where the eggs or roe of the female, and the milt of the male are contiguously deposited on the rocks or in the gravel at the bottom. The positive electricity of the warm water derived from the frictional action of sunlight upon the rocks and sand on the bottom of the shallow waters in which the eggs of the fish have been deposited, as well as upon the eggs themselves coming in contact with the negatively electrified eggs and milt evolves heat, and with it magnetism, and in due time the young fry are fully developed, vivified by these elements of life, breaking the outer membrane or shell of the eggs containing them, already distended and thinned by the growth of the embryo within, emerging into full life into the element where they are to have their being. Of course, the hatching of the eggs of fish is not uniform as to time in different species, some requiring a longer period than others to attain the maturity of their development.

Here we have a remarkable illustration of the production of

life by electricity and magnetism, outside of the bodies of the parent fish; while perhaps in almost every other class of animal life it is developed within the body of the female, after impregnation by the male animal, showing most conclusively that these imponderables are always present as well at the commencement of life as during its continuance, while it has been demonstrated time and again, that whatever decreases the *vis vitæ* of an animal diminishes also the evidence of the electricity within it, until after death it ceases altogether. Are we not right, therefore, in concluding that electricity, magnetism, and heat are, in certain relations to each other, elements of every life?

Oxygen gas is a supporter of combustion, as it also is of life, which in fact is one form of combustion. It is negatively electrified, and it is because it is so electrified that it supports both life and combustion. Let us illustrate this. The atmosphere, composed of nitrogen and oxygen gases for the most part, with a slight admixture of other gases and watery vapour, which last contains a large portion of oxygen gas, is negatively electrified. Wood, coal, and vegetable substances, in a dry state, are positively electrified. Now when we have on our hearths wood as fuel, and from the condition of the wood as well as that of the atmosphere the combustion of the wood is slow and sluggish, we apply a pair of bellows to hasten it the common explanation of this use of the bellows is, that it brings more oxygen gas into contact with the slightly kindled wood than the atmosphere naturally furnishes, and hence the combustion is quickened. This is true, but it also brings associated with the oxygen gas its negative electricity, which coming into union with the positive electricity of the fire and the wood already slightly heated, produces increased heat, which the additional oxygen gas thus supplied nourishes into flame, and the fire is properly kindled. Potassium thrown into a vessel of oxygen gas, bursts into the most brilliant flame from the same cause, the potassium being positively electrified in a high degree and so it is, but in a lesser degree, with the other metalloids.

In regard to the non-producing and non-conducting powers of electricity by fatty or oleaginous substances, a very remarkable fact has been developed in relation to the human family.

It has for a long time been observed that in countries where the sugar cane has been cultivated, and where sugar has been manufactured from its expressed juice, the negroes employed in making it grow enormously fat from the unrestricted use of the warm juice of the expressed cane during the process of boiling. From this food, like the whale, they become surrounded by an envelop of fat, as do also the interior organs of their bodies. This fat is anti-frictional and prevents the evolution of electricity, which in the absence of the fat would be developed. Hence these labourers could no longer be procreative, and as their labour was very exhausting, the necessity for a new gang of labourers every four or five years became established on sugar plantations. This fact, in sugar producing countries, has kept alive and continued the negro slave trade to this day—and where it has been abolished and the coolie trade substituted for it, the same results obtain. No women are sent to the plantations with the coolies, for they become like negroes, virtually emasculated by the absence of their electricity. So that we may attribute to the loss of electricity in the producers of sugar the great obstacle to the abolition of slavery for so long a time in the British West Indies, and at the present moment in the Spanish Islands, in Brazil, and elsewhere as it exists.

The same deteriorating influences upon their organization from fatness, in other portions of the human race, appear in various parts of the world, preventing the development of their electricity and magnetism, by which their animal functions are impaired, and their intellectual faculties greatly weakened. The Esquimaux, Fins, Laps, and all inhabitants of high northern climates, requiring a fatty and carbonaceous food, are examples of this character. The inference to be drawn from this remarkable fact is that such persons as are opposed to an increase of population, and who resist the injunction to the Patriarchs of "going forth, multiplying and replenishing the earth," should select for their companions in life the fattest persons of the opposite sex that they can find, and they will be rewarded by an immense reduction in their household and educational expenses when compared with those of their neighbours who chance to be of a lean kind.

In connection with this subject of continuing a species of animal, I may mention that in Europe, as well as in this country, a very mistaken notion exists as to the best age at which young cattle should be propagated. The prevailing idea is that heifers should not be allowed to bear their offspring before they are four years old, and in the state of Pennsylvania they are not taxable before they have attained that age. Now, this is a fallacy, as I have abundantly tested during the last twenty years. I have thought that nature was the best guide in such cases, and accordingly, as my animals are always well cared for, my heifers are sufficiently developed and matured when nine months old to receive the masculine impregnation, and to undergo, afterwards, a healthy gestation, and to produce their young when about eighteen months old. By my system of breeding, there is a saving in the expense of supporting young heifers during two years and a half over the common method. My herd of cows thus produced will compare favorably in size, produce of milk, cream and butter, and healthfulness with any herd of similar numbers of cows in this country. I do not remember to have had a sick cow or heifer during the last twenty years. But I have exceeded even this early propagation of their species. Last year a young heifer of mine, only four months old, manifesting a desire for copulation, was permitted to receive the male impregnation. She duly conceived, and before she was fourteen months old she bore a healthy male calf. The heifer herself, apparently, was not incommoded by the event, and continued to enjoy excellent health; and some six weeks after the birth of her calf she again received the male impregnation. This heifer was reared under the stimulating influence of the associated blue and plain glass, which had hastened its development three years and a half. Now, apply this discovery to the rearing of domestic animals throughout the world, and begin to estimate the benefit to mankind to be derived from the reduced expenses in producing them and the great gain that will result in increasing the number of animals to be raised in any given period of time, and some faint idea may be formed of the great value of this discovery in this single branch of human industry.

A wide-spread error in agriculture exists in Europe, as well as in this country, and has even been maintained in books of science. It is "that underneath large trees vegetation droops and languishes, even when the shade is not very intense." Some years ago I had occasion to plough up the sod which covered a small orchard of apple and chestnut trees on my farm. All the trees were old and large. I caused the field to be well manured, even to the bottom of the trunks of all the trees. When the ground was well broken up, I directed my farmer to mark out drills for sugar beets, and to plant the seed close up to the trunks of all the trees. He looked at me with astonishment, and said: "Why, sir, plant so close to the trees? Nothing ever grows under the shade of trees!" I replied that I had heard such a statement before, but that I did not think it to be well founded. I had seen too many weeds, suckers and brambles growing luxuriantly under trees all over the country to attach any credence to it. "Do as I tell you; plant the seed close to the trees, and leave the result to take care of itself." My farmer was so much astounded by what he considered my foolish directions, that he went over to some farmers who were planting their seed in neighbouring fields, and told them of the absurd directions I had given him. In the fulness of their neighbourly kindness, they came over to me to enlighten me on the subject of farming. "Your man tells us," said one of them to me, "that you have told him to plant sugar beet seed close to the trunks of your big chestnut trees. We have come over to tell you, what you may not know, that no plant will grow under the shade of trees, and to dissuade you from attempting to make them grow there. We have been farming 25 years, and our fathers before us all their lives, and we have never heard of such a thing as planting for a crop under the shade of trees. Pray don't try it." I thanked them for their solicitude, but told them that "it was an experiment; if it should fail, the loss of a few seed and a little labour were all that would be involved in it; and if it should succeed, it would explode and banish a very mischievous and expensive fallacy in agriculture; little harm was to be apprehended from it." The farmer finding me determined, said, "You gentlemen from the city, come into the country, buy land, erect expensive buildings, purchase high priced stock of all kinds, and every new fangled tool or labour saving machine that is advertised, hire people and go to work, and think you are farmers; but I have never known one of you to make even his expenses out of his farming. You had all much better do as your neighbours do than strike out into new paths." I said to him, "your rebuke is just, and what you say is no doubt true; I acknowledge it to be true in my case. I know very little of anything, but I could not think for a moment of taking up the time of my farming neighbours by asking them how to manage my farm; I must learn it as best I can without taxing their neighbourly kindness, and this experiment of mine is one of my early lessons in farming." Finally, these good pe e took their leave, and my beet seed were planted accordingly to my directions. In due time they germinated, and began to grow, and to the surprise of my farmer the plants as they grew became stronger and larger at the bottom of the trunks of the largest trees than the other plants were in the open spaces in other parts of the field. This difference continued to increase as the season advanced, and when the time had arrived for gathering them, the greatest contrast was perceptible between those that had grown under the shade of the trees, even of the largest, and those which had grown in the open sunlight.

At this time the same kind neighbours who had visited me in the previous spring to advise me against planting my seed under the shade of the trees, were gathering their autumn crops in the adjacent fields. I went over to them and asked them if they would like to see my beet crop, and on their expressing a desire to see it, I invited them to accompany me, and we proceeded to the field. On our way I asked them where they thought the best beets would be found. "In the open sunlight to be sure," was the answer; "nothing ever grows under the shade of trees!" I made no reply, and soon after we entered the field. As we passed along I was amused at the astonishment depicted on their countenances as they examined the beets in different parts of the field. Presently one of them, nudging another, said in a low voice; "George, did you ever see any thing like that before? why, there are no beets in the sunlight, and the big ones are under the trees." This was the fact; the plants in the sunlight were few, scattered and spindling in their growth, having a long slender taproot and were valueless for food, while there was a luxuriant growth under the trees of large sized and excellent quality. After examining attentively the whole field, and declaring that they had never seen or heard of the like, and would not have believed it had they not seen it themselves, they came to me and asked me if I could explain so unheard of a phenomenon. I replied, "you know I am from the city, how then can I be expected to know anything about farming? If you who have been farmers all your lives, and your fathers before you the same, cannot explain this why should you expect me who have no experience in farming, being from the city, to do it? I know nothing about it, but I will tell you what I think. I will illustrate my meaning by an example: suppose you should take two men, both healthy, strong and vigorous, and both very hungry—one of them is six feet tall and very broad and muscular—the other man is five feet six inches high, and also muscular. Suppose you place them at a table and put before them food sufficient only for one man of average size and strength, and tell them to eat, how much of the food; do you think the little man would get?" "Well, I guess not a great deal of it," said one of the men; to which the others assented. "Now, suppose you had put on the table enough food for both, would they not rise from the table refreshed and reinvigorated, and ready for their work?" I said to them. "Well, yes, I should think so;" was their answer. "Now," said I to them; "the first supposition illustrates your mode of farming. You manure your land lightly, furnishing food enough only for your crop, and nothing for your hungry trees; if you should happen to have any upon your land. The trees, neglected and hungry, take all the food within reach of their roots, and nothing grows, therefore, under their shade—hence your proverb that plants will not grow underneath the shade of large trees even when it is not very intense. In my experiment I had placed sufficient food before the large trees, and the small plants. The tree digests its food, and can take no more food at a given time than can any animal, relatively—consequently what is left over after feeding the tree goes to feed the small plants and it also gets its fill of nutrition, so that both thrive and grow healthfully. Now, there is another reason why small plants should grow better and faster under the shade of large trees than anywhere else, and it is this. The dew late in the afternoon begins to settle upon the leaves of plants under the shade of trees an hour or more before it does out in the sunlight, and in the morning after the sun has risen, the shade of the trees protects the plants under them from losing the dew upon them by evaporation till ten o'clock, A. M. So that the plants under the shade of the trees have the advantage of four or more hours of moisture, in the dew that rests upon them, than other plants in the sunlight which have no such protection—and you know that moisture is necessary to the growth of plants." They thanked me for my explanation and went their way confounded. Since then I have cultivated under very large trees on my lawn, plants and flowers of many discriptions with great success, and the cultivation has greatly benefited the trees themselves. I would recommend to all having trees on their lawns to cultivate the soil at their bases in flowering plants, if they desire ornaments, or in vegetables if they need them for food. To holders of small patches of land, this information may prove to be of great comfort and convenience.

This little narrative brings me to the subject of the formation of dew, which I do not attribute to condensation of the atmosphere holding it in suspension, but to the exactly opposite cause, viz: the expansion and rarefaction of the atmosphere by heat, its ascent upwards and its abandonment of the water which it had previously held in suspension.

When, in the rotation of the earth upon its axis, any given area of its surface is no longer illuminated by the sun's rays, or, as in common language, it is said, "It is sunset;" the rays of sunlight do not illumine the atmosphere that is over such an area of the earth's surface, and, as the night advances, that atmosphere becomes colder and more magnetic with its increase of cold by induction. Columns or volumes of this cold air are then attracted to the earth by its opposite magnetism, and descend towards it. At the same time the air in contact with and just above the earth's surface, having been heated during the day by the electricity evolved by sunlight, and being positively electrified, ascends to meet the cold air descending from above, negatively electrified and oppositely magnetic; the conjunction of these opposite electricities produces additional heat which so warms the air freighted with moisture that is descending from above, that its expansion and rarefaction will no longer admit of its holding in suspension the watery vapour that it was bringing down with it; it consequently ascends alone, leaving the globules of water which it contained to be carried to the earth by their magnetism, and to insensibly settle upon the grass, leaves, earth, &c., and form what we call dew, hoar frost, &c., according to the temperature of the earth's surface at the time of such deposition. This occurs in a cloudless sky.

When the clouds are floating above us, there is no dew, not because, as we have been taught, that the radiated heat from the earth is reflected by the lower surface of the clouds to the earth, thus keeping the air in contact with the earth too warm to deposit its water as dew, as that is an absurdity, since heat reaching the lower part of any gaseous or vapoury fluid, would at once penetrate and permeate such gases, vapours or clouds and expand, rarefy and disperse them; but because the interposing clouds would prevent the descent of the volumes of cold air freighted with moisture above them to the earth below, and consequently there could be no deposition of water or dew from them. Cold does not condense the atmosphere, for if it did the density of the air would be much greater in winter than in summer, which we know is not the case. Besides, the rarity and tenuity of the air at great elevations, where extreme cold prevails perennially, contradicts this assumption. Nor has the air any weight—gravitation is supposed to act only in one direction, viz: towards the centre of the earth, while it is known that the air presses equally in all directions, upwards from below, laterally and downward from above, hence it cannot be acted upon by gravitation. The barometric pressure of the atmosphere in its variations, is due in all probability to magnetic attraction and repulsion between the atmosphere and the earth. The same reasoning applies to the waters of the oceans. They are fluids pressing like the air in all directions, upwards from below, laterally and downwards, and rest upon the earth by the attraction of the earth's magnetism, and not by gravitation, since their upward and lateral pressures are antagonistic to the attraction of gravitation. Every drop of water is a magnet. When the globules are vertical their poles are at the foci of their forms, the lower pole attracted by the magnetism of the air above and its upper pole attracted towards the magnetism of the earth below. These downward and upward attractions and corresponding repulsions dislocate, from their great mobility, other globules of the water, and force their polar magnetic axis to be horizontal or dia-magnetic, and these pressures everywhere varying in tension, develop magnetic forces throughout the mass of water, acting at every possible angle with each other, and producing everywhere opposite resistances. These magnetic changes induce electrical disturbances in the water, resulting in the development of heat by friction and the conjunction of opposite electricities, causing in all latitudes those currents of evaporation associated with electricity, which we find agglomerated in the atmosphere as masses of clouds, fogs, mists, &c. These masses of clouds acquiring their electricities by induction, become oppositely electrified according to their elevation in the atmosphere above the earth, and as they approach each other in their movements, an electric discharge takes place, a decomposition of the watery vapour occurs, the hydrogen gas is burnt in the oxygen gas of the decomposed water, displaying that bright yellow light peculiar to hydrogen, in flashes so dazzling that if they were not so evanescent no animal vision could support their glare and then follow their zigzag path in the atmosphere, as they are attracted by currents of oxygen in the air of varying conducting powers. The result is water electrified and magnetic, the globules of which repelling each other, and pressed upon in every direction by the magnetic forces of the atmosphere, descend to the earth as spherical drops to meet and mingle with the magnetism of the earth. These drops of water are what we call rain.

If it were not for the upward pressure of the waters of the ocean from their lowest depth, how long would the crust of earth beneath them, (computed by physicists to be relatively to the mass of the earth no thicker than an egg shell is when compared to the mass of albumen that it contains,) be able to sustain the pressure downwards of a mass of water from five to ten miles in depth as it moves in its tides, its currents, and the rotation of the earth upon its axis, and as it rolls in its orbit? Would not the momentum of such a mass of waters thus put in motion, in the course of time that has elapsed since they were gathered in seas and oceans, wear away so much of the earth's crust as to allow the waters to flood the interior fires of the earth, and produce explosions that would shiver the planet into thousands of fragments? And does not this furnish another argument against the doctrine of gravitation? The same principle applies relative to the upward pressure of the atmosphere. In the cases of the waters of the ocean and the atmosphere—both being fluids, differing however in their tenuity, their molecules have great mobility among themselves respectively, and from the irregular and unequal upward and downward magnetic attractions and repulsions, these molecules are displaced and turned aside, changing the directions of their poles and their axes, and thus becoming dia-magnetic or horizontally magnetic, creating thus the lateral pressures existing both in the water and the atmosphere.

When, from the mobility of the molecules in the crust of the earth at the period of the planet being launched into space in its rotary motion on its axes, and its progressive motion in its orbit, the equatorial diameter was, by magnetic attraction and repulsion, increased twenty-six miles more than the polar diameter, the same influences repelled from the poles respectively and attracted to the respective opposite poles the waters in the arctic and antarctic basins till they met in the tropics.

The upward pressure of these waters, their polar currents of cold water at great depths, and the rotation of the earth on its axis from west to east, have united in forcing the masses of oceanic waters to the westward till they impinged upon the eastern coasts of America and of Asia—action and re-action being equal; these waters, after their impact with these coasts and their contiguous islands, were reflected back again towards the western coasts of Europe and Africa, and meeting midway in oceans, the succeeding waves of these waters have risen above the general level of the oceans a few feet, which has been called a tide, and which has been attributed erroneously to the attraction of the sun and moon instead of to the forces which I have mentioned above.

The impact of these waters in mid-ocean throws back to the European and African waters, coming from thence and to eastern American and Asiatic coasts, the waters attracted there by the rotary motion of the earth on its axis—and thus they force back in all these continents the waters of the rivers emptying themselves into the oceans, creating in them the tides, the causes of which never before have been satisfactorily explained. These tides, therefore, are the results of the magnetic attraction and repulsion of the waters and the coasts of the continents where they are seen and felt—and are not affected at all, either by sun or moon.

The currents of the Mediterranean sea—the upper one inwards is the result of the pressure of the Atlantic ocean in its reflux from the mid ocean impact of the oceanic waters, the lower current running into the Atlantic ocean—is produced by the upward pressure of the Mediterranean waters and the magnetic attraction of the colder polar current at great depth towards the equator.

The heat of the earth ascends perpendicularly to the horizon. It cannot, therefore, be deflected to any considerable extent in producing winds or currents of air. These result from electrical and magnetic attractions and repulsions—the upward pressure of the air, which is nothing more than the magnetic repulsion of it from the earth—having their similar poles of magnetism adjacent, until by induction the polarity of the air is changed in the higher atmosphere, where, being intensely cold, it is attenuated by the repellent qualities of its homogeneous magnetism, and not by the low degree of its temperature, which happens to be coincident with its magnetism, but is incapable of condensing the molecules of the atmosphere.

When we remember the law of attraction and repulsion of

magnetism, viz: that it acts inversely as the square of the distance, and that the earth, its oceans and its atmosphere, are all magnetic, and mutually attract and repel each other according to this law—which, by the way, is the same law that Newton assigned to the gravity of matter—and when we further remember that they are all in contiguity with each other, we cannot fail to conceive that this planet has all the forces within and around it that are necessary for the performance of all its functions without attributing them to the actions of such distant orbs as the sun and the moon. If the moon, as our astronomers assert, exerts a greater influence upon the tides than does the sun, owing to the greater distance of the sun from the earth, by a parity of reasoning, how much more influential must the earth itself be which is in contact both with its waters and its atmosphere. All fluids when acted upon by unequal forces assume a spiral course, as witness the whirlwind in the atmosphere, and the whirlpool, and eddying currents in the waters. The currents of the oceans are spiral curves modified in their curvatures by the fixed as well as movable obstacles they encounter in their several courses.

When a wave at sea has reached its crest, why does it curl over and break into spray, as it descends into the trough of the sea? If the moon lifts it up why does not the moon hold it up? When a wave breaks on the shore, why does it cling to the earth, and recede in contact with it as the undertow, frequently carrying with it to destruction the incautious or unskilful swimmer? Why does not the moon keep this water on the surface instead of suffering it, though it be warmer than the water at greater depths, to seek its company against an assumed law of physics, that the warmer fluid floats upon the colder?

Why, in the whirlpool, does the warm surface water rush down its spiral coils to meet and mingle with the colder water of the greater depths? And why does this cold water ascend in counter spirals to meet the descending warmer water? This action is not caused by gravitation; it is magnetic, and so it is also in the whirlwind. The warm air of the lower atmosphere, in contact with the earth, is taken up in its spiral coils, attracted by the opposite magnetism of the upper air, which descends in opposite spiral coils to meet it in its ascent, and together the column of whirling air, repelled from its source and carried over the surface of the earth, but in contact with it, with a resistless impetuosity, by the electrical current which has developed the magnetism of the column, devastates and destroys every obstacle that lies in its course, till the magnetic equilibrium is again attained, when a calm ensues. In these instances of the whirlpool and the whirlwind, the assumed law of gravitation is violated by the ascent of the warm air into the colder upper atmosphere, as well as by the descent of the warm surface water to the depths below; thus proving that the motions of fluids, whether gaseous or liquid, are controlled by magnetism.

A balloon charged with hydrogen gas, and released from its fastening to the earth, ascends rapidly into the upper atmosphere—the region of intense cold, where, as we are taught in the schools, it should be condensed, and the sides of the balloon should be loose and pressed inward by the condensing power of the cold in that elevated region. According to the doctrine of gravitation it has ascended because it was filled with hydrogen gas—the lightest substance in nature—and every light substance floats upon any other substance heavier than itself.

Now, let us see what actually takes place in the balloon.

First, The hydrogen gas is positively electrified, and is attracted to the upper atmosphere by its opposite electricity, which is negative.

Second, The balloon itself is painted and varnished with gums to retain the hydrogen gas, which pigments and varnish are also positively electrified and assist in raising the balloon.

Third, The higher the balloon ascends the greater is the attraction of the negative electricity of the upper air for it.

Presently a conjunction of these opposite electricities of the upper air and the positively electrified gummed surface of the balloon occurs, heat and magnetism are evolved, the canvas of the balloon begins to expand, and within it the hydrogen gas also expands to fill and to tighten the canvas. The attraction from without and the expansion of the hydrogen gas within distend the canvas to its fullest extent. Should the æronaut not at once open the safety valve of the balloon, and liberate a portion of the hydrogen gas within it, these forces would burst the canvas and precipitate the unlucky æronaut to the earth, a catastrophe which really happened in England only a few days since.

The ascent of the balloon, the expansion of its canvas and of the hydrogen gas within it instead of their condensation by the extreme cold of the upper atmosphere, the bursting of the balloon—all contradict the Newtonian theory.

We will now explain why the temperature on the surface of the earth is greater during summer, though the sun is then at its greatest distance from the earth, than it is in winter, when the distance between the earth and the sun is at the least, being three millions of miles less than it was at the summer solstice—viz: June 21st. On this day the rays of sunlight, vertical at the tropic of Cancer, impinging through the atmosphere upon the surface of the earth, with a velocity of 186,000 miles per second, produce great friction. This friction is the result of the impact of all the rays of sunlight upon the earth's surface. This friction evolves more electricity in the contact than it does in winter, when the angle of incidence of the rays of light is very much more acute, and a large portion of the rays of light are at that time reflected and refracted into planetary space, without developing the electricity either in quantity or tension, which the whole quantity of rays of light would do if they reached the earth directly. Consequently as the electricity evolved is less in winter, the heat which this electricity produces in conjunction with the opposite electricity of the earth's surface is much less, and the temperature is therefore lower in winter than in summer.

Besides, the vertical impact of matter upon matter, as of light upon the atmosphere, or upon the surface of the earth, is always more violent, and produces more friction than its impact from an acute angle, or as it is called a "glancing blow," would do, hence more electricity results from the friction produced by the vertical impact of light, than there would be from its impact at an acute angle. The declination of the sun, therefore, by constantly changing the angles of incidence of its light, as it enters our atmosphere, and impinges upon the earth's surface, is the cause of the changes of the terrestrial temperature at the several seasons of the year. Hence the more vertical the light, the more friction is developed in its impact with the earth, and the more electricity thus evolved, and the more heat produced by the conjunction of the opposite electricities from the light and earth.

At the height of five miles or more above the earth, when masses of clouds oppositely electrified come together, great heat is evolved by the union of these electricities, and with it is also developed magnetism; the air of the cloud thus heated becomes positively electrified, and greatly expanded by the heat, it rushes upwards attracted by the negative electricity of the atmosphere above it, abandoning the watery vapour it had contained in suspension, and which absorbing the magnetism developed by the union of the opposite electricities begins to fall towards the earth, not by gravitation but by the magnetic repulsion of the surrounding air, and the magnetic attraction of the earth itself and the waters on its surface. At the same time, when this conjunction of opposite electricities occurs, much of the watery vapour that the clouds held in suspension is decomposed by the superior attraction of the intense electricity for the hydrogen gas of the water, which is immediately burnt in the oxygen gas that had been liberated by the decomposition of the watery particles of the clouds in the first place. This inflamed hydrogen burning with a yellow light, rushes to embrace again its lover, oxygen gas, pursuing it in those brilliantly illuminated zig-zag courses which we call flashes of lightning.

Now as these conjunctions of opposite electricities are successive in a storm, we see the frequent flashes of lightning and hear the rolling of the thunder, (which latter is merely the noise of the explosions of oxygen and hydrogen gases, when acted upon by a current of electricity passing through them,) as they dart or roll through the atmosphere. The water thus formed, starting in sheets or columns as it may be, is at once disintegrated, by the repulsion of the magnetism which it has absorbed, into atoms or globules, each of which is a separate magnet. These are repelled by the magnetism of the upper atmosphere, and are attracted by the opposite magnetism of the earth and its waters, and continue to descend towards the earth, but the molecules of atmospheric air are also magnets, and repel and retard the descent of the rain drops as they fall, and these forces continue to diminish their sizes, till, on approaching the earth, they are so comminuted, that frequently they become absorbed by the atmosphere and appear as mist and fog.

Now, if rain falls by gravitation, beginning, at that great height of five or more miles, to descend in the first second of time 16.1 feet, in the next 32.2 feet, in third second 64.4 feet, in the fourth second 96.6 feet, increasing its velocity as the time of descent and the space through which it passed as the square of the time, it would be found that its velocity and momentum, when it reached the earth, would be so great as to wash the soil into the seas, denuding mountains and disintegrating rocks, and destroying every living object on the planet. We see on a small scale the devastating power of a waterspout that breaks and discharges its contents when traveling only a short distance above the earth. Besides it is only necessary to see the retardatory effect of magnetism upon the flakes of snow as they fall lazily to the earth, each crystal of the snow flake, or frozen water, being acknowledged as a magnet endowed with its full proportion of magnetic power.

These facts prove that neither the clouds that float in the atmosphere nor the waters they contain, which have been taken up by evaporation from the rivers, lakes and seas, and which are again returned to them in rain, snow and hail, are affected by the so-called laws of gravitation. Conceive for a moment that the volume of water of the Niagara river which passes over the falls, should, by gravitation, descend from a height of two, three or five miles above the earth, the common height of clouds; then imagine the destruction that would follow such a descent; and yet water from clouds start in their courses towards the earth in masses so great as to dwindle in comparison the mighty stream of Niagara at the falls, and yet only benefit results from the rainfall. Why, then, does the water from the clouds not continue to fall, as it has started, in these enormous masses? It is because the Creator has beneficently provided against such a calamity by investing water with magnetism, when its constituents, oxygen and hydrogen gases, are combined by the passage of a current of electricity through them, in the formation of water, and the atoms or globules of water, being each magnetic, repel each other, and are repelled from the upper atmosphere—also magnetic—and are attracted to the earth by its opposite magnetism, allowing rain, snow and hail to fall gently and in small particles to the earth. Hence the greater the height of the clouds from which the rain falls, the smaller and more attenuated will be the rain drops in arriving at the earth. Mists and fogs, therefore, are as frequently the results of rain falling from very high clouds, as they are from evaporation at the surface of the earth or ocean.

Melted lead on the top of a shot tower is positively electrified—the air around it negatively electrified. The lead in falling repels itself and is attracted by the opposite electricity of the air, causing it to separate and to assume the spherical form of shot on reaching the vessels to receive it at the bottom of the tower. So that we may attribute the spherical or spheroidal forms of rain drops, of meteors, and of the planets themselves, to the forces of magnetism.

Let us take a cast iron spherical shot of the calibre of twenty-four pounds, and heat it to a nearly white heat; then let us select the lightest down from the common thistle that we can find; we will then shake some handfuls of it over the hot shot at the distance of three feet above it. It will be found that notwithstanding what is called the attraction of gravitation, not only of the heavy shot but also of the still heavier earth on which it is supported, the down will be carried upwards into the atmosphere by the current of heated air radiated from the hot surface of the shot, instead of falling either upon it or on the earth immediately adjacent to it. If, therefore, this heated shot repels some of the lighest flocculent matter of which we have any knowledge, and will not allow it to fall upon the earth in opposition to the radiating power of its heat, what becomes of the gravitation of the earth and of the other planets, and of cometary matter, &c., to the sun, if this latter is an incandescent body of a temperature so high that we cannot really conceive of its actual intensity? If the lightest substance, so-called, cannot be attracted by it through such excessive radiation of its heat, how can it attract the heaviest planets? What also becomes of its magnetism in the presence of such intensity of heat? It is evident that this great heat could not co-exist with the magnetic forces of the sun, which are thought to control the movements of our solar system.

Let us observe a boy on an August day, when the thermometer indicates 98° of Fahrenheit, in a room with closed doors and window sashes so as to admit no disturbing currents of air, while he amuses himself with blowing soap bubbles from the bowl of a clay pipe. When the bubble is formed, and it is sufficiently thin, he throws it off from the bowl of his pipe. The circumference of the bubble interrupted by the bowl of the pipe, as soon as it is detached therefrom, closes upon itself by magnetic attraction, and forms a nearly perfect sphere, while it ascends rapidly towards the ceiling of the room. Mark the play of iridescent colours on its surface as it receives the light from a window, just as the sun receives the separate rays of light from the stars and reflects them to the earth, &c. Now why does this bubble ascend in the atmosphere? The water and the soap of the bubble, as well as the component parts of the soap are each heavier than the warm air of the room. The gas that fills its interior, composed of vapour and carbonic acid gas from the lungs of the boy, is also in its components heavier than the same air, and is also probably of a lower temperature than the air, which is 98° of Fahrenheit, and yet the bubble, in defiance of the so-called laws of gravitation, ascends to the ceiling, instead of descending to the floor.

If what astronomers tell us is correct, the density of the sun is about one-fourth of that of the earth, and cannot relatively be so great, volume for volume, as that of this soap bubble. Water is the standard measure of density; potash and soda in salts, component parts of this soap bubble, have each a greater density than water, while the oil associated with them in the soapy water is perhaps less than that of water, while the density of the soapy water is greater than that of the sun. Now the earth, with all its power of alleged gravitation, could not prevent this soap bubble from ascending in the air. Now why was this? The globules of soapy water were held together in the bubble by the viscous character of its oily particles, which having an opposite electric condition to that of the water, attracted it to complete the circumference of the bubble when it was detached from the bowl of the pipe, while the magnetism of the whole bubble, repelled by that of the earth, caused it to ascend into the upper air by the attraction of the magnetism existing there.

Now conceive of a soap bubble 1,400,000 times greater in its dimensions than the earth, to be placed in one of the foci of the earth's orbit, and then imagine it to exert its gravitating power upon the earth, and estimate the result. If the earth could not attract by gravitation this soap bubble in the room referred to, what power would the big soap bubble have to attract the earth by its gravitation, when their positions would be reversed?

The undulatory theory of light is faulty in this, that every wave requires a resisting medium to lift it above the common level. In water, when any force disturbs its surface, the inertia of the water, against which the surface water is driven, offers a resistance by which the surface water is raised into a wave, but in all such cases the velocity of the force is small;

when the velocity of the wind, for instance, is one hundred and fifty miles per hour, it carries off the surface water into spray, until sufficient time has elapsed to allow the inertia of the mass of water to resist the impulse of the wind, when waves are formed. Now if the ether of interplanetary and interstellar spaces furnished such a medium of resistance it would not admit of the passage of light through it, with its inconceivable velocity of 186,000 miles per second. If the ether itself was luminous, some force of very low velocity must impinge upon it to make its undulations, and to be undulations they must meet with resistance to become such; besides all undulations occur on the surfaces of fluids, and extend but a short distance below the surfaces; but ether of space has no dimensions, it is illimitable; no one can say where is its surface; neither words nor figures can define its depth, width or height, and as all motions through it are of inconceivably high velocities, it follows that there can be no undulations in it, as they are produced by low velocities.

Sunlight, on a bright July day, falling in its greatest intensity upon the calm and placid surface of an expanse of water, penetrates it and descends to very great depths below it, without producing the slightest undulation on its surface, or movement within its masses. Its velocity is so great that no appreciable time is afforded for the disturbance of the inertia of the water. So it is with the ether of interstellar and interplanetary space. Thin, subtle, and attenuated, as this ether may be supposed to be, the velocity of light in passing through it is so transcendently great that there is no time for the disturbance of its inertia, and consequently its motion is instantly absorbed by the mass of the ether, without producing any undulation whatever. Now undulation is a superficial act. There is no wave at sea of a greater depth below the surface than forty feet; all below that depth is unaffected by whatever cause that may have produced the superficial wave. The great Leviathan of the deep, ninety or one hundred feet long and of other corresponding dimensions, plunges beneath the surface of the ocean when struck by a harpoon, and with inconceivable speed rushes into the depths below, yet he leaves no wave, no ripple, to indicate the course he has taken, and the whalemen in his pursuit have to scan the horizon in every direction to ascertain the place, sometimes a great distance off, where he has risen to the surface of the ocean to blow off his surplus electricity and carbonic acid gas generated in his lungs. So it is with all the fishes and marine animals that inhabit the great deep. Their motions, however slow or swift, develop no undulations beneath the surface, and consequently none appear on the surface; there are, therefore, no undulations below a depth of forty feet from the surface.

Geographers inform us that three-fourths of the outer crust of the earth are covered by water, only one-fourth being dry land. Of this fourth part but a small portion is habitable by animals, and a still smaller part thereof is actually occupied by them, while the waters of the earth are teeming everywhere with animal life. Innumerable myriads of fishes, marine animals, and sea monsters are known to exist beneath the surface of these waters; their speed in pursuing or avoiding each other, as they rush madly through them, should greatly disturb their even surfaces, but whatever agitations may occur in the depths of the ocean from these causes, no trace of them ever is seen on its surface; there is no undulation from such causes. Why? The reason is obvious. Fluids press equally in all directions. The inertia of the great mass of waters is not to be disturbed by the passage of even innumerable objects of small dimensions at whatever speed they may attain. The same principle obtains in relation to the ether of planetary space. This planet rolling in its orbit with a velocity of sixty-eight thousand miles per hour, through this ether, does not and cannot disturb the inertia of the whole ether of space; the motion of the part displaced by the earth and its atmosphere is absorbed at once by the whole mass, and its inertia remains unaffected; and so it is with all the planets, and even the sun itself. The sun's motion in its orbit being 14,400 miles per hour, the moon advancing in her orbit at the rate of 65,000 miles per hour, and so on with the rest of the planets, their enormous velocities will not admit of the disturbance of the inertia of the ether of space before the planet has left the ether far behind through which it has passed. The retardation of cometary matter in its course is not due to the resistance of the ether through which it is passing, for if it was it would be uniformly and continuously retarded in its whole course, and not merely as it is approaching or leaving the neighborhood of the sun, but it is owing to the magnetism of the sun and the planets, as well as of the opposite magnetism of the ether acting upon its own magnetism, that such variation in its velocity has been observed. This reminds me, that when a planet is at its nearest point to the sun, it is moving with its greatest rapidity in its orbit; and when at its remotest point from the sun, it is proceeding at its slowest rate of speed in its orbit; but yet the orbit throughout its entire course is so balanced that the rapidity is exactly proportional to the nearness, and the slowness to the distance in reference to each, so that equal areas of the space included in the orbit are described by the planet in equal times, which is Kepler's celebrated second law.

The friction of the atmosphere with the ether in its passage through it evolves negative electricity, which is taken up by the atmosphere by induction, and thus it becomes negatively electrified. If the planets cannot, in their rotation around the sun and on their respective axes, disturb the ether of space in its inertia, how can it be supposed that rays of light passing through it with its velocity of 186,000 miles per second, can cause it to undulate? Time is an element in the production of a wave, and in the passage of light through ether there is not time enough to resist the passage of light, in order to produce it. A musket ball with the initial velocity of 1500 feet per second, when shot from a musket will perforate a door hanging on its hinges without moving it, as there is not furnished sufficient time to disturb its inertia before the ball had passed through the door. So in like manner a tallow candle discharged from a musket will pass through a door without disturbing its position, while if it should be thrown from the hand against the door at the distance of ten feet from it, its momentum at such low velocity would push the door back to its frame.

Rays of sunlight, in passing through the ether of space, carry with them the negative electricity with which they were repelled from the sun's photosphere, and continue to be repelled by the negative electricity of the intensely cold ether itself through which they are passing. Now interpose a glass prism to the passage of a beam of this sunlight after it has reached us on the surface of the earth. This white beam of light is then refracted and decomposed, and each colour leaves the prism, diverging not only from the original ray of white light of which they are the elements, *but also from each other*, as may be seen by observing the spectrum which they form. This spectrum exhibits these colours in the order of their susceptibility of refraction, the red being refracted least and the violet most. From its appearance, Sir Isaac Newton, who first analyzed it, thought that there were actually seven primary or distinct colours in the composition of light, but since his day investigation and analysis have determined that there are but three primary colours, viz:

red, yellow and blue, and that the orange, green, indigo and violet, result from a commingling of the primary colours in different degrees of intensity, as they form the spectrum. Now, let us see what causes this refraction and decomposition of light by the prism. The glass prism was positively electrified when the sunbeam was thrown upon it; the opposite electricities of the light and the glass were brought into contact; heat and magnetism were evolved by their union; the glass was expanded by the heat, which was immediately absorbed by the air; the rays of light, changing their electricities by induction, become positively electrified and magnetic and repel each other, forming Newton's seven primary rays, according to the different degrees of positive electrization and magnetization they have absorbed. This explanation will also account for the invisible heat rays outside of the spectrum, which by some philosophers have been erroneously supposed to have come directly from the sun, associated with its light. Again, let us take two pieces of flannel made of wool, of the same texture and size; let one of them be white flannel, the other black flannel. Now white flannel has the same electrical condition as white sunlight, that is, negative. It consequently reflects or repels the sunlight, according to electrical laws. For this effect it is extensively used by the people of hot countries for articles of outside clothing to keep them cool during sunshine. Suppose we place these two pieces of flannel, in the winter time, on the snow, one hundred feet apart, the temperature of the air being at zero of Fahrenheit, and the sun shining brilliantly through a clear atmosphere, and let us watch the effect. In a little while it will be seen that the piece of white flannel is frozen tight to the snow, while the black flannel, having absorbed all the rays of the sunlight from its opposite electrical condition, has become heated by the development of the heat from the union of these opposite electricites, and the snow has become melted under the black flannel. This experiment proves that heat is the result of the union of opposite electricities as in the associated primary rays of light, for the material composing the two pieces of flannel was similar, while the negatively electrified white flannel repelled the negative white sunlight, absorbing the cold of the snow beneath and becoming frozen to it, as the positively electrified black flannel attracted the negatively electrified white sunlight developing the heat which melted the snow. Now as every object in nature has a colour of some kind, when the sunlight falls upon it, we can understand that the variations of temperature on the surface of the earth, are the immediate results of electrical action upon it by the rays of light as light and not by rays of heat from the sun.

We have thus shown you, that from the attributes of heat, it is physically impossible for it to be transmitted to this or any other planet from the sun through an almost infinite space of ether at a temperature of —142° of centigrade thermometer.

We have shown you that the negative electricity of our atmosphere is derived by induction from this very cold ether in the rotation of the earth on its axis, and in its motions in its orbit, carrying with it its atmosphere in its course.

We have shown you that the atmosphere is held in its place around the earth by its magnetism and dia-magnetism, which have been developed by currents of opposite electricities in conjunction, produced by the passage of rays of light through the atmosphere, evolving by their friction with it electricity of one kind, while the opposite kind of electricity has been produced by the impact of rays of light upon the more solid parts of the earth's crust and upon its waters as it developed their evaporation.

We have shown that the attraction of matter on or above the earth, is through magnetism to the poles opposite respectively to the hemispheres of the earth, that it is confined to the crust of the earth, and that it is not the attraction of gravitation.

We have shown that the upward pressure of all fluids, from capillary attraction in tubes to the upward pressure of the waters of the ocean that float the tonnage of the world, to that of the atmosphere which holds it suspended above the surface of the earth, is strictly magnetic. We have shown that the variations of the barometer at the level of the sea are not occasioned by the varying weight of the atmosphere, but by its magnetic condition, as those of the thermometer are produced by currents of electricity, which permeate the glass tubes that contain the thermometric fluid.

We have shown that all terrestrial heat is derived from the conjunction of opposite electricities, whether proceeding from the combustion of inflammable substances, from friction, or from the contact of currents of air or of gases oppositely electrified.

We have shown that friction of substances of low temperatures produces negative electricity, and increases the cold by their union, illustrated by two blocks of ice rubbed together and uniting more firmly at their junction than in any other of their parts. And then we have shown that positive electricity is always associated with heat, and the opposite electricity with cold; that their conjunction produces heat or cold according as one or the other of the electricities predominates at the moment of their union; that magnetism is also evolved by their conjunction, and that if much heat is developed, the magnetism disappears and takes refuge in the nearest greater cold; that magnetism is therefore the antagonist of heat, and is found in its greatest intensity in extreme cold, in the highest part of the atmosphere, and in the Arctic and Antarctic regions.

If the atomic theory be true, and the atoms of ether be spheres or oblate spheroids, we may imagine that light passing in rays through the intensely cold ether, develops negative electricity by its friction with the ether, and that this negative electricity resides in the interstitial spaces between the atoms of the ether until attracted by positive electricity of greater or lesser volume and tension, their conjunction would produce magnetism which would find a habitat among these interstitial spaces of the atoms of ether in the poles of the atoms themselves.

From the mobility of the particles of fluids, whether liquid or gaseous, it appears that their tendency is to move in spiral curves. In the currents of ocean, sea, lake or river waters, the frequency of their curved direction is everywhere manifest, any obstruction to the general direction of their currents, whether superficial, or at varying depths below the surface, is sufficient to determine them into spiral curves of greater or lesser curvatures. It would seem that this attribute of fluids was intended by the Creator for the evolution of currents of electricity by the friction of these particles of the inner curves of the spirals, and of magnetism by the passage of this electricity along the spirals of the fluids themselves. This is an origin of magnetism, as well in the waters as in the atmosphere. The great currents of the ocean, sweeping in curves greater than a great circle of the earth itself, are only elements of immense spirals. The circular motion of an infusion of tea in a cup when stirred by a spoon to hasten the solution of the accompanying sugar, is but an illustration of the same principle, and so it is with gaseous fluids. The tiny whirlwind that raises the dust in summer in our country roads, is but a type of the currents of atmospheric air, from the gentle breeze that fans us in the summer heats to the tornado, hurricane, and mighty cyclone that desolate the oceans and islands in intertropical regions. This form, therefore, in which these fluids are continually moving, is among the means adopted by the Creator to develop electricity, magnetism and heat, on and above the surface of our planet.

"Let us for a moment consider the action of the two great currents of warm water on the opposite coasts of North America. The Gulf Stream and the Japanese current through Behring's Straits to the Arctic Ocean. Let us consider the Gulf Stream. On the Equator, in the Atlantic Ocean the mean temperature of the surface of the sea, according to Kämtz, is 78.6°, the average maximum in latitude 6° north is 80.3°, the highest observed temperature in 3° 1′, north, according to Kotzebue, 84.6°, and the mean temperature of the sea between the parallels of 3° north and 3° south, according to Humboldt, was from 80.1° to 82.4°. The mean temperature of the air in the equatorial belt of the Atlantic Ocean between 10° north and 10° south, according to Lentz, is 78.8°. Here you have the surface water of the ocean in the Equatorial belt of the Atlantic Ocean hotter by 3.8° than air just above it. Now, if these respective temperatures were produced by emanations of heat from the sun, their condition of temperature should be reversed, the capacity of the air to absorb heat being so much greater than that of water. This fact proves that it is not solar heat that produces the temperature either in the air or water.

"In July, the course of the Gulf Stream, in latitude 38° north, shows the form of a tongue of temperature of 81.5°, (at some places even 84° was observed.) This hot stream produces itself as a double tongue, with a mean temperature of from 77° to 81.5° of Fahrenheit, (20° to 22° of Reaumur,) towards the north as far as the 40° of latitude, and towards the east to the 43° of longitude west of Greenwich, that is, far beyond Newfoundland. In January, the tongue of 77° of Fahrenheit, (20° of Reaumur,) reaches to latitude 37° north and longitude 70° 30′ west, and at the place where the east end of this tongue of 77° of Fahrenheit terminates in July, we find in January a temperature of 62.5° and 62.8° of Fahrenheit, (14° and 15° of Reaumur.)

"Up to the meridian of the eastern end of Newfoundland, the Gulf Stream proceeds first in an east northeast, and then in an east direction parallel to the American coast, with an average temperature in July of 77° to 83.8° Fahrenheit, (20° to 23° Reaumur,) and in January, of 68° to 77° Fahrenheit, (16° to 20° Reaumur.) The highest temperature of the air in Africa in the same parallel of latitude in January, is only 59°.

"At Newfoundland, the Gulf Stream comes in violent collision with the Polar Stream of Labrador, which nearly at a right angle sets against and penetrates into it like an immense wedge. On the eastern side of the Grand Bank it is so powerful that, according to the surface isotherms, it penetrates into the Gulf Stream from 150 to 200 miles southward of its general limits, and therefore entirely intersects the surface waters of the easterly stream for that breadth, which is the most important part of its course. The Gulf Stream, 300 miles northeast of Newfoundland bank, after having passed beyond this polar current, is *warmer* than it is south of it. The influence of the temperature of this polar steam is less in January than in July. 380 miles eastward of Newfoundland, on the 50° of north latitude, the Gulf Stream has a surface temperature of 68° Fahrenheit in July, while in January, the Gulf Stream on the 50° degree of north latitude has a temperature of 54.5° Fahrenheit; the thermometer shows at the same time at Prague, or at Ratibor, (in Silesia,) on the same parallel of latitude, temperatures of minus 24°, and sometimes still lower ones. The isothermal line of 54.5° Fahrenheit, (10° of Reaumur,) runs up in July towards Iceland and the Faroe Islands to the 61° of north latitude. There it meets for the second time the polar stream which on the east coast of Iceland again threatens to block up its way and to destroy it. In July, temperatures were observed on the north coast of Iceland of 45°, 47° and 49.3°, (by Lord Dufferin, 46°,) while off the east coast for six degrees of longitude, none higher than from 40° to 42.6° were found.

"According to Irminger's data, and Lord Dufferin's observations, the Gulf Stream setting towards the north preponderates in July on the north and west coasts of Iceland, but on the east and south coasts the polar stream coming from the direction of Jan Mayen.

"Between Iceland and the Faroe Islands, the Gulf and polar streams are contending against each other, and the result of this struggle is a sea divided into a great number of hot and cold bands, which fact is demonstrated clearly by Lord Dufferin's cruise from Stornoway to Reikiavik in 1856, and fully corroborated by Wallich in the Bull Dog Expedition of 1860.

"The fact that the two streams in their contest appear as many bands and strata alongside, over and beneath each other, is proved not only by the observations of the temperature of the surface of the sea by Irminger and Dufferin, but also by the researches of Wallich in regard to the nature of the bottom of the sea. The latter found there volcanic stones pointing as to their origin to Jan Mayen, and at other places ophiocomæ of two to five inches in length which could have been carried there only by the warm Gulf Stream. Besides, the drift ice penetrates here further to the south than anywhere else east of Iceland. * * * * But here the Gulf Stream comes away equally intact from its struggle with the polar stream as at Newfoundland. We now know its further course in the summer from many direct observations as far north as Spitzbergen and Nova Zembla, and beyond the 80° of north latitude.

"The mild winter of the British Isles is well known. The mean temperature for January in London is 37.4°; at Edinburgh the same; at Dublin 40.5°. The further we go from east to west or from south to north, or, in other words, the nearer to the Gulf Stream, the higher we find the temperature. At Unst, on one of the Shetland Islands, 560 miles north from London, the mean temperature of the air in January is 40.3°, and that of the sea 45.5°, (East Yell.) The warm current of the sea is tempering the air. The lowest temperature observed in London was —5°, at Penzance on the west coast, +24.1°, at Sandwick on the Orkney Islands +15.8°, at Madrid +13.3° has been observed, and +27.5° at Algiers, which provides Europe with cauliflowers in winter.

"On the morning of Feb. 8, 1870, the telegraph announced the temperature at Ratibor, (in Silesia,) to be —25.4°, while northwest of it, at Breslau, it was —13°, at Berlin —0.4°, at Kiel +10.6°, and at Christiansand, on the south of Norway, 8° of latitude north of Ratibor, + 30.7°. So high a temperature would be impossible in Norway if the winds did not bring it from the high temperature of the Gulf Stream to the westward.

"Many persons suppose because the summer in Iceland is rough and cold that the winter must be dreadful in its severity of cold, but exactly the contrary is the case. Dr. Henderson states, that 'I really shuddered at the thought of living through the winter in Iceland. How greatly was I astonished when I found the temperature not only higher than in Denmark, where I had been during the preceding winter, but also that the winter in Iceland was by no means more severe than the mildest winter which I had ever known in Denmark and Sweden.' Sheep and horses have to take care of themselves during the entire year in Iceland; only cattle and the more valuable saddle horses are fed in the stable during winter. How impossible would it be in Germany to leave any domestic animal in midwinter without shelter even for a few days only. The lakes near Reikiavik, in Iceland, are frozen in many winters not more than two inches thick, very rarely to eighteen inches. The lowest temperature of the air experienced there during thirteen years was only + 3.9°.

"It is not to be wondered at that such is the case, because the warm Gulf Stream provides Iceland with heat. Its mean temperature there is, even in January, 34.7° above zero, and the *lowest temperature noted during twenty years was only* 28.6°. Iceland is situated close to the Arctic circle, and in the latitude of Siberia.

"While on the western side of the north Atlantic ocean, the polar ice reaches down to latitude 36° north, (the parallel of Gibraltar and Malta,) and the name Labrador is sufficient to characterize the climatic qualities of all the land between 50° and 60° north, there exists on the east side of the ocean along the Norwegian coast cultivated land up to 71° north, the northernmost land of the world, in which, under the influence of the Gulf Stream, agriculture is the main occupation of the inhabitants. Wheat is grown up to Inderoen, in latitude 64° north; barley up to Alten, in 70° north, where sowing generally is done between the 20th and 25th of June, yielding in the short space of eight weeks, to the 20th or 30th of August, in the average six or seven fold; the potato yields at the same place on the average seven or eight fold, in favourable seasons even twelve to fifteen fold; it thrives on the coast as far east as Vadso, on the Russian boundary line. At Alten (70° north) relishable cauliflower is raised even in less favourable summers. Where washed by the polar current, there are, as shown by the various Franklin expeditions, under 70° north, but desolate ice deserts without any cultivation. There is on the eastern side of the ocean the flourishing and busy little town of Hammerfest, where only once the temperature has been as low as +5° and generally is not less than 9.5°, while on the western side of the ocean there are only the poor snow huts of the Esquimaux in 70° north.

While Germany has to suffer the frigid air of —24°, and sometimes more intense cold in winter, at that same time Norway gathers a rich harvest under the Arctic circle, not from its acres, but in the warm waters of the Gulf Stream, as for instance at Ansvaër, in the direction of the vortex of the Gulf Stream; there the herring makes its appearance about the 10th day of December, remaining until the first days of January, and then about 10,000 people congregate, and haul about 200,000 tons of these fish of a value of more than one million of dollars."

The warmer air of the land near large bodies of water, whether of lakes, seas or oceans, is due to the difference of temperatures between that of the atmosphere and that of the waters, which being in contact at the surface develops one kind of electricity, which meeting with the opposite electricity of the air evolves heat and renders the climate of such localities mild, healthful and agreeable.

"East of the North Cape, distant from it about 120 nautical miles at Vardöe, the temperature of January is +18.5°; while at St. Petersburg, 620 miles south of the former, it is +15.1°, or 3.4° colder. But the most important fact, testifying to the existence and the great volume of the Gulf Stream at the North Cape, appears to me to be the temperature of the sea at Fruholm, which in January is in the mean still +37.9°. Fruholm is on the same parallel of latitude as Ust-Jansk, latitude 70° 55′ north, in Siberia, and Point Barrow, in North America. The former has a mean temperature in January, of —38.6°, the latter of —18.6°. Meran, in Tyrol, of world wide celebrity, on account of its mild and temperate air, nearer to the equator by 24½°, has in January a temperature of the air of 31.8°, Venice, 36.3°, Vevay, 33.1°, Paris 35.4°, New York, 29.5°, Washington, 31.5°."

We will not pursue this subject of the surface temperature of the Gulf Stream to its ultimate northern development, but we will turn our attention to the temperature of the Gulf Stream, at its various depths in its course, as well as of the sea itself.

"North of the isothermal line of 39.4°, (3.3° of Reaumur,) toward the pole, the temperature generally increases with the depth, while southward, toward the equator, it decreases. There is, however, no uniformity in this, as Lieutenant Rodgers, in 1855, found in the Asiatic part of the Arctic Ocean there is on the surface a warm current, with water of a low specific gravity, beneath it a cold current, and then again a warm current of heavier water, and all these strata running in opposite directions.

"In entering upon the question of temperature of sea water at different depths, it must be borne in mind that water is densest at a temperature of 39.2°, and that it arranges itself in the various depths according to the specific gravity in strata, either above and beneath, or alongside each other. From the place where the sea shows at the surface a temperature of 39.2°, it will lose in temperature toward the pole, while in general, it will gain with the increase of depth, but toward the equator the temperature of the surface will increase while it will decrease downward in proportion.

"Parry, in latitude 57° 51′ north, longitude 41° 05′ west of Greenwich, on June 13th, 1819, observed the sea to have a temperature on the surface of 40.5°, and at a depth of 1410 feet, in the Gulf Stream, 130 nautical miles southeast of Cape Farewell, a temperature of 39°. 140 miles northeast of this place, in latitude 59° 35′ north, longitude 38° 5′ west of Greenwich, Captain Kundsen, on the 30th of June, 1859, found the temperature of the surface 44.6°, and at the depth of 1800 feet, 43.4°, which corresponds with Parry's measurements.

"Wallick remarks that on the parallel of latitude 63° north, not far from the south coast of Iceland, the temperatures on the surface, and at a depth of 600 feet, differ in the average not more than 3.8°, and that consequently the Gulf Stream does not essentially lose in temperature to that depth.

"On Irminger's chart of the currents and ice drifts around Iceland, there is, in Brede Bugt, (Broad Bay,) in latitude 65° 17′ north, longitude 23° 25′ west of Greenwich, a temperature recorded of 46° at the surface, and of 45.5° at a depth of 300 feet, showing that the Gulf Stream at this place in the vicinity of the Polar Circle has lost in that depth only .5 of a degree of temperature.

"Scoresby remarks, 'that the temperature of the sea near Spitzbergen is six or seven degrees warmer at the depth of from 600 feet to 1200 feet than it is at the surface.'

"From the results obtained by the British Sounding Expedition, from May 31st to September 7th, 1869, in the North Atlantic Ocean, between the Faroe Islands and Spain, it appears that the Gulf Stream has, between Ireland and Spain, a depth of 900 fathoms or 5400 feet, and equally as much near the Rockall rock, west of the Hebrides. Between Rockall and the Faroe Islands, near the parallel of latitude 60° north, it reaches to the bottom of the sea, which has a depth there of 767 fathoms, or 4602 feet, and at that depth the Gulf Stream has still a temperature of 41.5°. It has also been found that an *Antarctic* current of cold water, directly over the bottom of the sea clear up to the Irish and Scottish coasts, exists, meeting there an Arctic stream. In the notes of Professor Thomson, the stratum at Rockall, from 900 to 1400 fathoms below the surface, is designated as cold indraught, Arctic and Antarctic, "(temperature 39.2° to 37.4°,) and the stratum between 900 and 2435 fathoms, between Ireland and Spain, as indraught" of cold water, probably mainly Antarctic, (temperature 39.2° to 36.5°.)

"It is demonstrated by figures and facts, that the hot source and core of the Gulf Stream extends from the straits of Florida along the North American coast at all times, day and night, in winter as in summer, even in January, with a temperature of 77° and more, up to the 37° of north latitude, while at the same time and in the same latitude in Tunis, in Africa, the temperature of the air is but 53.4°. The Gulf Stream transports and develops still in this latitude a higher temperature than either water or air possesses in the Atlantic ocean, even under the equator, on which neither in July nor in January the temperature is ever as high as that of the Gulf Stream in latitude 37° north.

"Under the 37° and 38° of northern latitude, the hot core of the Gulf Stream turns away from the American coast towards the east beyond the meridian of Newfoundland and its bank to 40° of longitude west of Greenwich, where it still possesses a temperature in July of about 75°, and in January of about 66°. From there it proceeds to the northeast, diffuses nearly across the entire Atlantic, and surrounds the whole of Europe to the Arctic region and the White Sea of Archangel, with a broad and permanent warm water course, without which England and Germany would be a second Labrador, and Scandinavia and Russia a second Greenland, buried beneath glaciers; whereas, in Fruholm, (71° 6′ north,) the sun does not rise at all above the horizon during the entire month of January, in a latitude in which, in Asia and America, the mercury remains frozen for months—there the Gulf Stream preserves for the sea a temperature of 37.8°. While the sun in the short days of winter sends forth his rays of light and warmth but for a few hours, and the influence of the latter is quickly lost again in the long nights, the Gulf Stream does not cease, day or night, to be the source of warmth.

"The Gulf Stream carries more heat to the north than is carried by all the warm air currents from the entire periphery of the equator towards the North Pole and towards the South Pole. The southwest winds receive their high temperature from the Gulf Stream, and only through the ocean—not by the winds—can warmth be carried into latitudes as high as those of the European coasts are.

"From the soundings obtained so far, the Gulf Stream must be, up to the Arctic ocean, a deep and voluminous water course. If it should not be so, the polar ice would reach also the European coasts. In the Antarctic ocean the polar ice drifts all around the globe as far at least as latitude 57° 5′ south, in many places to 50° and 40°, (latitudes corresponding respectively to those of the British Channel and the Mediterranean Sea,) on some even to 35°, (corresponding to the latitude of Morocco,) but not the smallest particle of northern polar ice has ever reached even the northernmost cape of Europe. The Gulf Stream in its course is more powerful and steady than all the winds; only the the polar ice and polar currents in spring and summer exercise a great influence over it. The polar stream presses at three places against it: first, from the northwest, east of Newfoundland, then from the northeast of Iceland; at both these places the polar stream is buried and proceeds beneath the Gulf Stream, after having pushed it off laterally to the southeast. But for the third time, at Bear Island, the polar stream comes directly against the Gulf Stream from the northeast, splits it into two or three branches, and in places even presses it beneath its own waters at least in July. Under the lee of Spitzbergen, this latter branch rises again and proceeds on the surface according to Parry's observations to latitude 82½° north. The main branch east of Bear Island, has been traced by Dr. Bessels to latitude 76° 8′ north, where in August, 1869, it had still a temperature of 41.2°.

"The polar streams, in conformity with the general laws of nature, are less powerful in winter than in the summer. The polar ice does not drift as far southward; it makes fast more or less to the Arctic coasts and islands; in spring and summer, on the contrary, it drifts along similar to the glacier tongues, in Alpine mountains, or the ice in our rivers. The Gulf Stream is in winter more powerful than in summer, while the polar streams, so to say, set at rest in some measure, withdraw their ice and concentrate it around the land. The relations of the temperature of the Gulf Stream within themselves, are about the same in January as in July, the fluctuation between its maximum and minimum temperature, (July and January, or August and February,) would be on the average only about 9° of Fahrenheit, (4° of Reaumur.)

"What immense contrast to this extraordinary temperature is offered by the temperature of the air on the mainland! From the sea and air isothermal line of 36.5° Fahrenheit, (2° of Reaumur,) at Philadelphia, to Northumberland Sound, with —40°, the distance is 2280 miles nearly due north, There is, therefore, in about each thirty miles a fall in temperature of one degree, as you go north. From the same point at Philadelphia to the Gulf Stream, east of Fruholm, on the same isothermal line of 36.5° Fahrenheit, (or 2° of Reaumur,) there are in the direction of the Gulf Stream, in an air line, about 5400 miles, in which distance there is no fall at all in the temperature of the Gulf Stream. There, one degree of fall in each thirty miles; here, the same temperature along 5400 miles in a northeast direction. Such is the influence and power of the Gulf Stream. In the latitude of Berlin, which has a mean temperature of the air in January of 28°, the Gulf Stream has 50°; at the Faroe Islands it has still 42.1°; but in Jakutsk, in the latitude of the Faroes, the air is 40° below zero, a difference of 82.1°."

Scoresby remarks: "In some situations near Spitzbergen, the warm water not only occupies the lower and mid regions of the sea, but also appears at the surface; in some instances, even among ice, the temperature of the sea at the surface has been as high as 36°, or 38°, when that of the air has been several degrees below freezing. This circumstance, however, has chiefly occurred near the meridians of 6° to 12° east of Greenwich, and we find from observations that the sea freezes less in these longitudes than in any other part of the Spitzbergen sea."

"The hot source and core of the Gulf Stream extends from the straits of Florida, along the North American coast at all times, day and night, in winter and summer, even in January, with a temperature of 77°, and more, up to the 37° of northern latitude, while at the same time, and in the same latitude, in Africa, (Tunis,) the temperature of the air is but 53.4°. The Gulf Stream transports and develops still, in this latitude, a higher temperature than water and air possess in the Atlantic ocean, even under the equator, on which neither in July nor in January, the temperature is ever as high as that of the Gulf Stream, in latitude 37° north."*

Why is this? We have shown that heat could not be forced down by the sun along the line of the Gulf Stream, by any power of which we have a notion. If this heat could be derived from the sun, it is clear that the temperature of the ocean under the equator should be at least as great, if not much greater, than it is in the straits of Florida, or up to the 37° of north latitude; but we know, experimentally, that this is not the case, but that the heat is actually less either on land or ocean under the equator, than it is in that portion of the Gulf Stream from the straits of Florida to the 37° of north latitude. Therefore solar radiation of heat is out of the question. Nor could the great heat at the immense depths of the Gulf Stream, penetrate thereto, even if it were possible for heat to descend to our planet from the sun, for the tendency of heat is everywhere to ascend into the atmosphere, and it could not remain permanently at those depths in opposition to that tendency. We must therefore seek the cause of this marvellous heat in the waters of the Gulf Stream, somewhere else than in the sun.

We are told by our geologists that very great heat exists in the interior of our earth—and the existence of volcanoes in many portions of the globe which are now active, as well as those which have been quiet for a period of time unknown to man, all attest the truth of their assertion. These volcanoes, past and present, have subterranean and submarine communications with each other, which permeate large portions of the interior of the earth and serve to transmit any excessive accumulation of heat from its immediate source to even the most distant parts of the earth's interior, for radiation to the surface of the earth. These communications are simply flues for distributing the interior heat of the earth to its various parts. The greatest heat is and always has been under the equator, and these flues are for the most part submarine. If you will

* From Dr. A. Peterman's Essays on the Extension of the Gulf Stream.

take an atlas of physical geography and cast your eyes upon the map showing the distribution of volcanoes and the regions subject to earthquakes, you will discover that the southern part of Mexico and the isthmus connecting the two Americas are studded with volcanoes, while the Caribbean sea is filled with them. These volcanoes are doubtless connected by flues which are united into many proximate flues in the straits of Florida, through which the surplus heat of the interior of the earth under the American continent and a part of the Atlantic ocean and the Gulf of Mexico is transmitted to the Arctic regions, warming the waters of the Gulf Stream through its whole length, and thus moderating the climates of the western parts of Europe. Another system of volcanoes will be observed almost on the same meridian, extending from Tristan d'Acunha in the southern Atlantic ocean though Trinidad, St. Helena, Ascension, Cape Verd Islands, Canary Islands, Azores, Iceland and Jan Mayen, to the Arctic regions. These volcanoes attest a central heat, forcing a passage by the repellent affinity of positive electricity with which it is associated in the direction of the polar axis of the earth, to outlets at either pole. When obstructions are met with in the passage of this heat and electricity towards the poles in the interior of the earth volcanoes are formed, the superincumbent crust of the earth is upheaved and a vertical flue or chimney instead of the original horizontal or inclined flue is developed, and an eruption of matter is thrown out to form an island, which in a series of ages may become a continent.

These two systems of submarine flues carrying the heat of the central portion of the interior of the earth under the Atlantic ocean, a part of the American continent, the Carribbean sea, Gulf of Mexico and the Antilles, meet under the Atlantic ocean to the southeast of the island of Iceland, each furnishing its supply of heat to maintain the temperature of the Gulf Stream, as well in its greatest depths as on its extended surface. As heat ascends from its source into the atmosphere, it passes upwards from the bottom of the Gulf Stream through it to its surface, associated with its positive electricity, where it encounters the negative electricity of the atmosphere, and by conjunction with it, increases the heat of the air above the water, which air, thus warmed, attracted by the colder air negatively electrified of the land that is nearest to it, flows in a steady wind towards it, ameliorating its climate and promoting the health and happiness of its inhabitants.

All warm currents of water, wherever they may be situated, have a similar origin in the heat developed in the interior of the earth. The islands of the Pacific ocean may be all regarded as volcanic. The western coasts of America from Cape Horn to their northern limits, furnish a corresponding proportion of volcanic action, and the warm Japanese current through Behring's straits and along the coast of Asia, evinces a similar origin in submarine flues conveying heated air under the ocean to the Arctic regions on that side of the globe.

"The British expeditions for deep sea soundings ascertained the temperature of the water of the Gulf Stream, at a depth of 6000 feet, (being more than one mile,) to be 38.1°, and at 14,610 feet, (being nearly three miles,) to be still 36.5°. Compared with this, the deep sea temperature of the Gulf of Arabia, and even of the water under the Equator, will be found very low, sinking to 34°; in general, the deep sea temperature of the tropical oceans is lower than that of the North American basin.

"In the northern Atlantic ocean, between 50° and 60° of latitude, there are certain bands of water of a high tempera ture interposed between bands of water of a lower tempertu re.

"*These bands of a higher temperature are to be found, more or less, where a warm current and a cold current converge, as, for instance, east of Iceland.* The two principal bands alluded to by Admiral Irminger, in his memoir, in about 60° of north latitude, between the Shetland islands and Cape Farewell, are, doubtless, the two convex vertices of the Gulf Stream in that region.

"The fact that the entire sea between Scotland and Iceland consists of a great number of such warm and cold bands of water, adjoining each other, is best proved by the cruise of Lord Dufferin, who, sailing from Stornoway, in the Hebrides, to Reikiavik, between the 13th and 20th of June, 1856, observed the temperature of the surface of the sea every two hours—in all, ninety times—and found it to change not less than forty-four times, or, in the average, once in fourteen nautical miles, the change fluctuating between 52.9° and 43°; for the most part, however, between 50° and 47.8°; while on starting from Stornoway, the temperature was observed to be 48°, and on arriving at Iceland again 48°.

"There are bands where the water is of a higher temperature close to one where it is of a lower temperature, and such bands are found on each passage across the Atlantic, between Fairhill and Greenland. The difference between the highest and the lowest temperatures of the sea observed on this line of the Atlantic ocean is 10.8°, up to 30° or 40° west of Greenwich; to the west of this meridian, the temperature fell more rapidly, the more so the nearer to Greenland. The temperature of the warmest bands is defined frequently pretty sharply against the waters which run through them. This high temperature of the sea at its surface, extends 30 degrees of longitude, or at least 900 nautical miles west of Fairhill.

"Findlay mentions that the temperature at the depth of 1200 feet was found to be only 55°, while on the surface of the Gulf Stream it reached 77.4°. In the Florida straits, where the velocity of the Gulf Stream is greatest, the temperature at 4800 feet was found to be only 38.1°.

"The warm water of the Gulf Stream is not found at considerable depths, much of the heat of the lower strata escaping to the surface. It is, besides, a fact, that this warm water is but little apt to mix with the adjoining sea-water.

"Above the broad Atlantic ocean, in high latitudes, in the colder seasons there is a relatively high temperature, which by the prevailing western and southwestern winds is carried to the coasts of Europe."

Let us now consider, some of the recognized laws of heat and electricity. It is known, that where two adjacent different temperatures exist there electricity is evolved. Now the waters of the Gulf Stream, the Japanese current, and of other hot streams existing in the oceans and along coasts, deriving their heat in the first place from the submarine flues connecting subterranean and submarine volcanoes with the Arctic and Antarctic regions, admit of the passage of this heat through their globules to their upper surfaces, in conformity to the attraction of heat from the surface of the earth to the upper atmosphere. This ascent of heat from the bottom of these hot streams through their waters to the atmosphere, in connection with the indraught of cold Arctic and Antarctic waters flowing over the bottom of the oceans, is the cause of the low temperature always found at such depths in those waters—while intermediately from the bottom of the ocean to the surface in such hot currents of water, the temperature varies till it comes into contact with that of the atmosphere, and that of the ocean water encompassing these hot currents of water through their whole extent. The contact of these different temperatures evolves electricity, which is positive where the high temperature of the water pervades its greater volumes, and negative electricity where the cold Arctic and Antarctic waters exceed in volume, below the surface, the waters of the hot stream. The conjunction of these opposite electricities evolves heat, which being absorbed by the water where they meet serves to supply a continuous source of heat to the farthest extremities of such hot currents of water to the Polar regions—and this is why this great heat is maintained from its original source in the Florida straits to the high latitude where it is observed. The cause of the hot waters of the Gulf Stream not mixing readily with the colder waters of the Northern Atlantic ocean, will be readily found in the junction of these opposite electricities, producing heat where these hot and cold waters meet.

In ascending from the earth in a balloon, aeronauts have discovered the same law to prevail among gaseous fluids as among liquid fluids on the earth, and that strata of heated air, even at great elevations, are as it were sandwiched between others of far lower temperature; the contiguity of these strata of warm and cold air develops heat and electricity as well as magnetism in the atmosphere, as is done also in the waters of the ocean by corresponding columns of warm and cold water in juxtaposition. These attributes of fluids are, therefore, among the great sources of the evolution of these imponderable powers.

The cold Arctic and Antarctic currents of water, in motion to the Equator from the poles while currents of warm water from the tropics to the poles are moving beside them in a directly opposite direction, are conclusive evidences that they are impelled by magnetic attractions and repulsions in the crust of the earth, and so it is also with the aerial currents of the atmosphere. Those of a great elevation, having a very low temperature, are attracted towards the Equator and downwards to the earth by its magnetism, while the warm equatorial currents, repelled from the earth by the same magnetism which has attracted the cold upper current downward towards it, ascend to the upper regions of the atmosphere attracted by the opposite magnetism existing there, and in both cases in opposition to the supposed law of gravitation, for the air descending to the earth from the elevated regions of the atmosphere is much thinner and more attenuated than the air beneath, and the ascending warm air is much denser than the air of the regions that it seeks. The diagonal and spiral motions of either the descending or the ascending currents of the atmosphere are produced by the magnetism of those portions of the atmosphere, through which they are respectively passing.

When our attention is directed to the fact of the Labrador and Polar, or Arctic currents running towards the Equator, while by their sides the Gulf Stream is running towards the Arctic regions in an opposite direction; and when it is discovered by the deep sea soundings, that there are currents of water of varying temperatures at great depths which also run side by side in opposite directions, at whatever depths, we are forced to the conclusion that no conceivable system of gravitation can be devised to explain the anomaly. But if we apply the law of development of heat and magnetism, by the conjunction of opposite electricities, which are always associated with differences of contiguous temperatures, the solution of the phenomena referred to becomes comparatively easy. The electro-magnetic condition of the warm water of the Gulf Stream is repelled from the Equator, and attracted by the opposite electro-magnetic condition of the waters and atmosphere about the North Pole, while the cold waters of the Labrador and Arctic currents are repelled by the similar electro-magnetism of the waters at their starting point, and are attracted towards the Equator by the opposite electro-magnetism of the warm waters there. Similar causes produce similar effects in the southern hemisphere, and similar electro-magnetic forces dominate in the atmosphere all over the planet. Hence we find there, horizontal winds blowing in opposite directions, one above the other, and it is by this wise arrangement of oppositely electrified currents of air that the rainfall is scattered and distributed over vast areas of the earth's surface, modifying the temperatures and furnishing to the parched and arid soil those supplies of water for irrigation, so indispensable to the support of animal and vegetable life upon it.

In the year 1828, I was detailed with two other officers of the army, by the Secretary of War, to make a survey of the mountainous region in the states of North and South Carolina, Georgia, and Tennessee, lying between the head of navigation on the Savannah river, at the eastern foot of the Blue Ridge mountains, and the head of navigation on the Tennessee river, on the western side of the same mountains. The object of the survey was to ascertain the practicability of constructing a navigable canal on the mountains, to bring the produce of northern Alabama and eastern Tennessee to Charleston, in South Carolina, and Savannah, in Georgia, instead of sending it to Mobile and New Orleans, and thus it was hoped by the administration of the Government to reconcile the people of South Carolina and Georgia especially, to the policy of having the internal improvements of the country to be made by the Federal Government instead of by the State Governments.

On reaching our destination, I was directed to run a line of levels from the head waters of the Savannah river over the mountains to those of the Tennessee river, a distance, if I remember rightly, of some ninety miles. I had under my command eleven men—mountaineers—stout, strong, active, and hardy fellows. The other officers were employed in prospecting for other routes across the mountains, at considerable distances from that I was pursuing. The country was then very thinly settled, and a portion of my route bordered on the lands occupied by the Creek or Cherokee Indians, then living in the state of Georgia. Of course, we had to carry all our supplies with us, the country furnishing little or nothing. We were occupied on this duty some five months, from July till December. Frost appeared in the latter part of September, on the parallel of latitude of Charleston, in South Carolina, and thin ice was formed on the streams almost nightly after October 15th. In the latter part of October my party was benighted in the valley of the Little Tennessee river, far away from any human habitation, on a narrow alluvial bottom, overhung by a precipitous and lofty mountain. The man detailed to bring to us from the mountain ridge our supplies for the day and night, had missed his way, and had descended to the river, at a place that we had left several miles behind us. He had not observed our trail, and supposing that we had not passed the spot which he had reached, he kindled a fire, and remained there all night awaiting our arrival. After sending men in every direction in search of him, who returned without success, I began to make arrangements for the night. The air was cold and humid, ice being formed of the thickness of a quarter of an inch on the still waters of a portion of the river, a heavy growth of timber in the valley of the river where I had halted rendered the ground, as well as the air, very damp. The men, like myself, were all dressed in light summer clothing, and fire, therefore, became a prime necessity, but the question was, how to obtain it. At that period, lucifer matches, if they had been invented, could not be procured where we were. My arms and ammunition, with the rest of our supplies, were with my wagon, and where it was we had not been able to discover. It occurred to me to procure fire by friction, for at that day it was thought that heat was evolved by friction. So I divided my ten men into five reliefs of two men each, and directing some of them to gather the driest pieces of wood they could find, I notched the pieces so as to make the greatest rubbing surfaces possible in them, and then I set two men at a time to rub the pieces of wood together. Having some pieces of dry paper in my pockets, I hoped to be able to kindle a fire with them, when sufficient heat should be developed by the friction of the pieces of wood. The men relieved each other every five minutes, after having rubbed the pieces of wood together, vigourously and rapidly; the wood became blackened, and much smoke was given out, but no fire could be produced. The wood itself was not sufficiently dry, and none more suitable could be procured. The evening air was cold and damp and carried off as fast as it was evolved the positive electricity which flowed from the friction produced on the wood by the active rubbing of the men. One of the elements therefore to develop the heat, viz: the negative electricity of the atmosphere that we needed, was wanting. After having kept these five reliefs of the men continually busy in rubbing these pieces of wood for two consecutive hours, I gave up the effort in despair, and we submitted ourselves to the circumstances of our situation, and passed a dismal night of great suffering. Had the wood and the night air been dry, we should have kindled a fire in fifteen minutes with such an amount of frictional electricity as was developed by the rubbing of the wood by the men. The experiment satisfied me that heat is only developed by the proper electrical conditions and not by friction of itself. As it was, all the friction we could produce did not prevent us from passing two days and nights in these mountains without food or fire, the water on the river, in its tranquil parts, having been frozen at night of the thickness of a quarter of a dollar or an English shilling.

Every housewife in the country knows that if she suffers the sunlight to fall upon the burning fuel on her hearth, the combustion of the fuel will be deadened by it, and if allowed to continue long, it will be extinguished. This is owing to the de-oxydizing power of the blue ray of the sunlight, which separating the oxygen gas from the atmospheric air in the chimney, prevents the combustion of the fuel from the absence of oxygen gas. Whoever has seen one of our western prairies on fire, must have observed, in the stillness of the morning air and in the bright sunshine, that the cumbustion of the dry grass and herbage was slow, the flame lazily creeping from one stalk to another till a canopy of smoke intercepting the sunlight, allowed a current of air to be formed beneath the smoke, which fanned the combustion into active flame. These results were from the removal of the oxygen gas from the air in the first place, by the blue ray of the sunlight de-oxydizing it, and in the second part, obscuring the sunlight by the canopy of smoke, which permitted the oxygen gas in the atmosphere to be re-united to the air beneath it, and to supply the oxygen gas to support anew the combustion on the prairie.

It is therefore a mistake to suppose that friction produces heat. It evolves electricity, which, uniting with opposite electricity, develops sometimes heat and sometimes cold, as one or other of the electricities is predominant in volume and tension at their conjunction. This is illustrated by the passage of sunlight through two adjacent panes of glass, one being blue, the other colourless and transparent, at the same angle of incidence. Glass is known to be a feeble conductor of heat as well as of electricity, for we use glass in our windows to confine within our rooms the artificial heat produced within them during winter, and in northern regions double sashes are used in the windows, the outer sash to prevent the cold from penetrating through them, and the inner sash to confine the warmer air within the rooms; and in electrical experiments, glass handles are used to insulate currents of electricity intended to be passed from one pole of the battery to the other.

Now when sunlight with its enormous velocity falls thus upon two such adjacent panes of glass, it will be found that the plain transparent glass is cold to the touch of the hand, while the blue glass is hot when so touched. If friction produced heat, both of these surfaces should have the same temperature, but such is not the case. The reason is obvious. The sunlight passes through the plain transparent glass, only slightly retarded by its density, which is greater than that of the atmosphere, but subject to its refraction—while six of the primary rays of the sunlight that impinges upon the blue glass, are suddenly arrested by the impact with it, which shatters the composite rays of indigo, violet and purple into their component parts, and only admits of the passage of the blue ray through it. This sudden stoppage of a velocity of 186,000 miles per second of six of these primary rays of sunlight produces enormous friction, which evolves negative electricity from these rays, which coming in contact with the vitreous or positive electricity of the glass evolves heat, that expanding the molecules of the glass allows the heat thus developed and a current of electro-magnetism, produced at the same time by this conjunction of opposite electricities, to pass through the glass, and to produce the marvelous results upon animal and vegetable life that we have announced. This, then, is the theory that explains the almost magical effects that are produced in life by the impact of sunlight upon the adjacent surfaces of plain transparent glass and blue glass.

The facts are in such harmony with the explanation of them, that as we cannot deny the facts we are bound to accept the theory that elucidates them. This will relieve the scientific mind that is always bothered to accept a new fact or to comprehend a new theory.

Light is diffusible. This is apparent everywhere in our illuminations. It is also compressible, as illustrated by the concentration of sunlight through a common lens or sun glass into a focus, by which a boy lights his segar or inflames a squib of gunpowder. This shows that rays of light move through ether, and our atmosphere, without touching each other, and that when they are compressed together, as in this lens, their tangency produces friction, and this friction evolves negative electricity, which has caused their separation, which negative electricity brought into contact with the vitreous or positive electricity of the glass of the lens, develops heat of extraordinary intensity. Now, when we come to apply these attributes of light to the physical condition of our planet, we are at no loss to assign the variations of our temperature throughout our seasons, directly to the action of light upon the various solid, liquid or gaseous constituents of the planet, which at certain times and in certain conditions are oppositely electrified to the rays of light.

There is no atmosphere about the moon and consequently

it has no heat, as the rays of light which fall upon the moon's surface being negatively electrified as they pass through the cold ether of stellar and planetary space, on reaching the moon at a very small angle of incidence from the sun, are instantly reflected from its surface upon the earth and into space. The moon itself being negatively electrified by its contact with this ether in its career in its orbit, this negatively electrified condition of the moon's surface repels the rays of light therefrom, and hastens their reflection. The rotation on its axis is the effect of electrical forces in its interior, and its motion around the earth, and with it around the sun, results from the magnetism contained within its crust, and in the earth and its atmosphere, as well as in the planets, the sun and the ether of space.

No one impulse could possibly send light from its various sources in the firmament through space with its constant velocity of 186,000 miles per second. It is impelled through space with its own concomitant forces, as a rocket fired from its stand is continually driven forward by the forces evolved in the combustion of its composition, till it is extinguished. So light is repelled from its sources in the firmament by its negative electricity, and its velocity is maintained by the assistance of the negative electricity of the ether through which it is passing, continually driving it forward. This condition of negative electricity in light being constant, and its velocity uniform, its rate of speed is maintained till it enters our atmosphere, where it encounters electrical disturbances of opposite as well as similar conditions, producing its refraction, its reflections, its polarization and its absorption. On reaching the surface of the earth, which at every moment presents a new portion to the action of light, all the phenomena of day, twilight and night, of heat and cold, of dryness and moisture, of atmospheric and climatic changes, are developed. Seasons succeed each other, according to the angles of incidence of the sun's light. When it falls in the summer on certain parts of the earth almost vertically, no rays of light are reflected from it, they all impinge upon it with their inconceivable velocity, developing by their friction with the earth an opposite electricity to their own and that of the atmosphere, whose union produces the heats of summer. In winter, though the earth is three millions of miles nearer the sun than it is in summer, yet the angle of incidence of the sun's rays of light is so small and acute, that a large proportion of them are reflected into space without producing the friction with the earth which is necessary to evolve an opposite electricity and heat consequent upon the union of the two electricities; hence the temperature of the winters in such parts of the earth's surface is low, and cold prevails. The intermediate seasons make an average between the extremes of summer and winter, from the corresponding angles of incidence of their light.

One of the most beautiful illustrations of the remarkable power developed by the compressibility of light is furnished in the celebrated exploits of Archimedes, the Syracusan, the most learned of the mathematicians of antiquity, in destroying by means of reflecting mirrors the fleet of the Romans, who, investing the city of Syracuse by land, were blockading its port with a numerous fleet, which was preparing to batter the sea walls of the city with battering rams and catapults. Archimedes conceived the idea of destroying this fleet, which was unapproachable by any adequatic force under the control of the Syracusans, by concentrating upon it the light of the sun, reflected from mirrors into foci, successively thrown upon the several ships of the fleet, at the distance of an arrow's flight from the shore, or from 150 to 200 feet.

The two ancient authors who have furnished the clearest account of this extraordinary feat in warfare, are Zonaras and Tzetzes, who each lived in the twelfth century of the Christian era. The passage in the history of Zonaras does not enlighten us in regard to the construction of the mirrors used by Archimedes, it simply states the fact, and in another passage the same author says, that under the empire of Anastasius, in the year 514, A. D., Proclus with burning mirrors burnt and destroyed the fleet of Vitalien, who was besieging Constantinople, and he added, their invention was ancient, and that Dion gave the honour of it to Archimedes, who had used it successfully against the Romans at the siege of Syracuse.

The historian Tzetzes, enters more fully into the description of the mirrors used by Archimedes, which he said were composed of a central hexagonal mirror, surrounded by others of a smaller size, which by the aid of hinges and metallic plates, could be so exposed to the sun, that its rays of light falling upon them would be reflected and then concentrated into a common focus, developing so great a heat that the ships of the Romans were burnt by it, even at the distance of an arrow's flight.

Among the moderns, Kircher has written that Archimedes had been able to burn, at a great distance, with plane mirrors, experience having taught him that in assembling in this manner the images of the sun, a heat could be produced at a oint where these images were united.

Mr. Du Fay, a member of the Royal Academy of Sciences, in a memoir printed in 1716, stated that the image of the sun, reflected by a plane mirror more than 600 feet, upon a concave mirror with a diameter of 17 inches, burned inflammable substances at the focus of this concave mirror. He moreover added that some authors had suggested that a mirror, with a very long focus, could be formed by using a large number of small plane mirrors, which might be held in the hands of as many persons, and so directed by them as to throw, by reflection, all the images of the sun upon a given point, thus developing great heat; but at the same time he treated the story of Archimedes burning the Roman fleet at Syracuse as the veriest fable, and worthy of all ridicule.

It is very singular that men will frequently believe statements of the most improbable and even impossible character, who, at the same time, will reject the best established historical facts when they happen to be outside their circle of knowledge. Such has been the fate of the history of the burning mirrors with which Archimedes destroyed the Roman fleet at Syracuse. This fact, related by many historians, believed, without question, during fifteen or sixteen centuries, was, in the seventeenth century, not only disputed, but was treated as a silly fable by many of the savans of that period. Even the illustrious Des Cartes openly denied its possibility, and we must acknowledge that with the then received opinions on Dioptrics, Des Cartes was excusable for not believing the mirrors of Archimedes ever to have existed.

This incredulity, on the part of many persons claiming to be scientists, excited the interest of M. de Buffon, the celebrated naturalist, at the time the Intendant of the Jardin des Plantes, at Paris. He determined to test the question practically, and for this purpose constructed a system of reflecting plane mirrors, by which he attained complete success. He began by measuring the loss of illuminating power in the reflection of the sun's rays from metallic mirrors of the finest polish, when compared with the loss so sustained by reflection from plane glass mirrors covered on their backs with tin foil. It was found that the glass mirrors lost less light by reflection than the metallic mirrors did, but that it required two plane glass mirrors of the same dimensions to produce, at a given distance, an illumination equal to that from the same unobstructed beam of sunlight passing into an obscure room through an aperture in the window shutter, and consequently, that the number of his glass mirrors should be largely increased to produce any sensible effect on combustible substances. After studying his subject in its various relations to the laws of light and heat, as then understood by scientific men, M. de Buffon constructed his mirror of 168 pieces of plane glass, covered on the back with tin foil, each piece being six inches wide by eight inches long, separated from each other by four lines, and mounted on a stand, which was susceptible of being moved in every direction; each of these glasses had a separate setting, so that it could be separately moved in every direction, independent of the movements of the other glasses. It required about half an hour to adjust the reflected images of the sun from these mirrors into a common focus. When the glasses were properly arranged, and the focus adjusted, a board of beech wood covered with pitch, was set on fire by 40 of these glasses at the distance of 66 feet; with 98 glasses, a board covered with pitch and sulphur was set on fire at the distance of 120 feet. A slight combustion was produced on a board covered with wool cut very fine, by employing 112 glasses, at the distance of 138 feet, with a very pale sun. At 150 feet of distance, a board covered with pitch was made to smoke with 154 glasses, and it was thought that it would have been burnt if the sun had not become overcast with clouds. With a still feebler sun, chips of pine wood covered with pitch have been set on fire in one minute and a half, at the same distance, with a like number of glasses. With an unclouded sun, a pine board, covered with pitch, at the same distance, has been quickly set on fire with 128 glasses, and the fire has caught the whole surface of the focus, which was 16 inches in diameter, at that distance. Finally, the focus having been shortened to the distance of 20 feet, with 12 glasses the substances easily combustible were set on fire. With 45 glasses a tin canister, weighing six pounds, has been quickly melted with 117 glasses. Thin scraps of silver have been melted, and a sheet of iron has been made red hot; and there was reason to believe that if all the glasses of the mirror had been used, metals could have been as easily melted at 50 feet distance as at 20 feet.

These experiments have been made with a sun of a spring time, and without much power, having been enfeebled by atmospheric vapours. If then, with these disadvantages, wood could be burnt at 150 feet distant, we may well think, that with a summer's sun, it could be readily burnt at 200 feet distance, and with three similar mirrors it could be set on fire at 400 feet distance. M. de Buffon thought that with mirrors similar to his own, combustibles could not be inflamed beyond a distance of 900 feet.

Let us attempt an explanation of these phenomena. The enormous velocity of rays of light in coming to our planet, establishes the fact that they cannot touch each other in their passage, since if they jostled each other their velocity would be greatly diminished. Repelled from each other, therefore, by their own negative electricity, as well as by that they have received from the cold ether through which they have passed, they are attracted to the glass of the mirrors and their metallic backing, by the vitreous or positive electricity of those substances. On striking the glass, these rays produce friction, which evolves positive electricity, the junction of these opposite electricities evolves heat and magnetism, the rays of heat thus developed follow the same laws as do those of light, and together, both are reflected from the mirrors and are directed to the common focus, where their concentration sets on fire combustible substances, and melts and vaporizes those of a more obdurate and intractable character. The refraction and reflection, as well as the polarization of light, are due to the repellent affinity of electricity.

When we are told that on many parts of the earth's surface mountains have been upheaved till their peaks and ridges, at distances varying from 16,000 to 28,000 feet above the level of the sea, appear to be covered with snow, which from year to year, and from century to century, continues to cover them, no matter in what latitudes they may exist, nor in what season of the year they may be examined, we naturally ask ourselves, why is this? How does it happen, that these snow-capped peaks and ridges, at such great elevations above the sea, far above the region of the atmosphere in which clouds and vapours habitually love to roam as it were at will, basking in a resplendent and brilliant sunlight, receiving all the supposed emanations of heat from the sun, that philosophers of every age have innocently conjectured that that luminary, like a human spendthrift, was lavishing upon infinite space, in all directions, that a small portion of it might reach our planet, should preserve their mantles of perpetual snow, in all seasons, in all climatic changes that are occurring every moment thousands of feet beneath them, and thus continue defying, as it would seem, the mutability of all other earthly things? Some of our philosophers of the highest distinction, have gone into the most elaborate calculations to show what enormous columns of ice, of the greatest density, could be melted by the heat of the sun, in its constant emanation, in the smallest spaces of time, in the face of the fact that the snow clad mountains, that happen to be the nearest to the sun, have been from time immemorial, unaffected in the slightest manner, by any heat derived from that great luminary. Let us attempt an explanation of this wonder. The colour of snow is white. It has a low temperature. Its electrical condition is negative, as is the white colour of sunlight, as are the rays of sunlight which reach us through the negatively electrified ether of space, also intensely cold, and the intensely cold upper strata of our atmosphere. As a consequence, white sunlight, negatively electrified, falling upon the white snow capped mountains, also negatively electrified, as are also the strata of our atmosphere into which these mountains lift their heads, these similar electricities repel each other. The white sunlight is reflected into space from the snow covered mountains, which remain undisturbed, and no trace of the action of heat, as derived from the sun, is anywhere visible upon them.

If the sun is a great magnet, it must have its magnetic poles, with their reciprocal attractions and repulsions. The plane of the sun's equator is said to be neither perpendicular to nor coincident with that of the ecliptic. Its magnetic poles may therefore be differently situated in it to the positions occupied in the earth by its magnetic poles. From the supposed enormous volume and intensity of magnetism in and about the sun, we may infer that the velocity of the planets and of cometary matter in their respective progress in their orbits, would be checked when in their several perigees or nearest points to the sun, from its great magnetic attraction, and that as they severally receded therefrom, those velocities would be increased from the loss of the sun's attraction by increase of distance from it, and the nearer approach to their apogees, or greatest distance from the sun, where the sun's attraction would be the least, and the opposite magnetic attraction of the ether of space would be the greatest. If it were not for the interior forces of the planets, &c., causing their rotations on their axes, we might suppose that their movements around the sun might be stopped entirely, when they had severally reached their perigees by the magnetism of the sun.

When two magnets of different magnetic volumes and intensities are brought near each other with similar poles towards each other, the greater magnet will repel the lesser : if their opposite poles approach each other, the feebler will be attracted by the stronger. Now the sun having much greater magnetic power than the earth, when the latter is at its perigee its velocity must be retarded by the greater attractive magnetism of the sun, which would hold it fixed when in perigee, but for the rotation of the earth on its axis, driving it forward, and that retardation or holding it back after it had passed its perigee would continue until the earth had receded so far from its perigee as to have reached the attraction of the opposite magnetism beyond its apogee.

The sun exhibits every characteristic and evidence of a body enveloped in two atmospheres, so to state, the one in contact with it being the region of white light, called the *photosphere*, and outside of that, a region in which coloured light is sometimes manifested, especially along the edges of the solar disc, and which last region is called the *chromosphere.* The spots on the sun are supposed to be holes of various forms and dimensions in the region of white light, through which the dark body of the sun itself has been seen. These spots or holes are liable to variations, and are analogous to the spots of sunlight on the surface of the earth, which are sometimes seen to be surrounded by the shadows cast upon the earth by the clouds above it. Nasmyth, in the year 1866, made the discovery that the luminous portion of the sun's disc is not composed of light of equal or homogeneous intensity, but consists of a minutely divided series of luminous streaks, which he described as like willow leaves, around which the light is less intense, or rather the photosphere is more transparent. These willow leaves appeared to cross each other in all varieties of directions, and their average magnitude was about one thousand miles long, by a hundred miles broad; other observers have preferred to describe these appearances as "granulations," "rice grains," and "shingle beach," and as having elliptical forms, and of much smaller proportions.

The moon, we know to be a reflector of light without the emission of any accompanying heat. The picture of the face of the moon exhibited to us, represents great irregularities in its surface, depressions, as if they were craters of extinct volcanoes, and elevations of great altitude, conveying the idea of volcanic mountains; but the general colour is that of a light grey, not unlike to sheets of zinc, or tin foil, the latter of which we use as backs or reflecting surfaces in our glass mirrors.

If we thus get our nocturnal light from the moon, unaccompanied by heat, why should we insist upon violating the well established laws of heat in its radiations, and declare the sun to be an incandescent body, continually in active combustion, requiring inconceivable masses of fuel of some kind to maintain it, and surrounded on all sides by an immensity of ethereal space of so low a temperature that any radiation of heat from the sun must necessarily be absorbed and neutralized as soon as it should leave the body of the sun? We therefore, for the reasons stated in this book, reject entirely the theory of the incandescence of the sun, and of its luminous metallic vapours of great intensity of heat.

We have shown in the body of this work, that the colored lights constituting the primary rays of light, which are emitted from the various orbs of the firmament, negatively electrified, and are propelled by the cold negatively electrified ether through which they are continually passing to the sun, and through its transparent or translucent chromosphere to the photosphere of the sun, are there commingled to produce its white light, which then is repelled or reflected from the grey "willow leaves," "granulations," "rice grains," or whatever they may be, into ethereal space by the same negative electricity, which has been associated with them throughout, a portion of which comes to us as the white light of the sun.

This shows the synthesis or formation of the white light of the sun, and that it is merely an association of the primary rays of light thrown together by electrical and magnetic attractions and repulsions in the photosphere of the sun, and so easily separable that the slightest change in the angle of incidence of the white light of the sun, as it falls upon vapours, clouds, or gases will excite their repellent affinities, and resolve them into the varied and brilliant tints of primary and composite colours, which everywhere in the temperate regions, serve to excite our astonishment, wonder, and delight. These changes need no accompaniment of heat, and as they are without it, we return to the declaration of Moses, that "God made two great lights, a greater light to rule the day, and a lesser light to rule the night and the stars.

"And he set them in the firmament of heaven to shine upon the earth, and to rule the day and the night, and to divide the light and the darkness; and God saw that it was good."

Among the fallacies of science, as taught in our schools, to some of which I have alluded in this book, there is not one more surprising than the statement made by our astronomers, that the earth, the planets, and the sun itself continually revolve on their respective axes, and in their orbits from west to east. We are also told that these orbits are elliptical curves which return into themselves. Now we will illustrate this movement by supposing that a man has started from San Francisco, on the Pacific Ocean, to travel on the same parallel of latitude from west to east around the world. After he has travelled one hundred and eighty degrees on this parallel of latitude, he finds that he has reached the east cardinal point from San Francisco, and if he should continue his journey, he must travel westward, which course will bring him in time back again to San Francisco. How is it possible, therefore, in a curve which returns to itself to travel always in the same direction? There can be no fixed cardinal points in any solar or stellar system which is always in motion. In regard to the diminutive planet which we inhabit, the curvature or annulus of magnetic poles, north and south, is sufficiently stable and fixed to furnish cardinal points of the compass to regulate our journeyings upon it; but with planets, stars, and suns, it is different. They have no fixed points in the celestial sphere, of which we have or can have any knowledge, to which the direction of their movements can be referred, and it is simply an absurdity to attempt to assimilate planetary and stellar motions to those of mankind on our earth.

The planes of the orbits of the planets are neither coincident with, parallel, nor perpendicular to each other, but they are supposed to intersect each other in such a manner that the sun shall always be in a focus, common to all of these elliptical orbits; consequently any perpendicular line or plane to any one of these orbits, cannot be perpendicular to any other of them; and hence, there can be no cardinal points common to them all, and their motions cannot be from west to east.

My task is finished. When, in the beginning of this century, it was announced that the primary rays of light had different attributes, and among them, that the blue ray stimulated vegetation in a remarkable degree, many persons on the continent of Europe, as well as in the British Isles, instituted experiments, with a view to utilize these rays. Their experiments were failures, as they were made with homogeneous tinted glass, each of the primary rays having in this way been somewhere tested, but without satisfactory results. A knowledge of these failures induced me to examine the subject of vegetable growth in its natural conditions. I soon discovered that where vegetation was most luxuriant, and exuberant, there the brilliant sunlight was always associated with the blue light of the firmament. That during the torpor of winter, the rays of sunlight fell upon the earth, owing to the declination of the sun, at such acute angles of incidence, that many of them were reflected into space without stimulating life on this planet, while, at the same time, the blue colour of the sky was intercepted from our vision by the watery vapours and clouds that were constantly floating in the atmosphere. The absence, therefore, of the blue colour of the sky, and many of the rays of sunlight at this season, together with its low temperature, convinced me that the Creator intended it to be a season of rest for vegetable and animal life, a sort of Sabbath, in which life, though existing in plants and animals, was reposing from its activity, to be aroused into exercise on the return of the season of spring, when from the less declination of the sun, more of its light would be thrown upon the earth, associated with the blue colour of the sky, then unmasked by the dissipation of the clouds and watery vapours which had concealed it during the winter just past. I said to myself, "here is the secret of the failures of these European experiments with the primary rays of light. I will follow the guidance of the Creator in cultivating my vines. I will associate the sunlight with the blue colour of the sky, intensifying the latter. I will make a tropical climate and atmosphere in the temperate zone," The results are before you. The reflections I have made on this subject have induced my investigation into the Physics of Nature. I have not been satisfied with what I have been taught in the schools. Their explanations are not consistent with the known or presumed facts. I have ventured, therefore, to form my own conclusions, irrespective of dogmas that have been thrust upon mankind for centuries. I do not profess to teach any one, but as a human atom among the masses of mankind, for whom all knowledge should be disseminated, I venture to impart to the public the conclusions to which I have arrived on these subjects, and that public may attach to them whatever value they please.

APPENDIX TO PART II.

[I.]

A very remarkable confirmation of my theory of the formation of the equatorial diameter of the earth, as well as of those of the other planets, by magnetic attraction and repulsion from their respective poles, thus increasing those diameters in various proportions over their several polar diameters, has unexpectedly appeared in a paper read before the American Academy of Sciences, at their meeting in this city held on Thursday last, November 4th, 1875, and sent to it by Professor Joseph Le Conte, of the University of California, a synopsis of which was published in the supplement to the *Public Ledger*, of this city, on Saturday, November 6th, 1875. The paper was entitled "On the Evidence of Horizontal Crushing in the Formation of the Coast Range of Mountains in California," being the result of recent observations by the author. His theory is, that mountains are formed wholly by a yielding of the crust of the earth along certain lines to horizontal pressure, not by bending into a convex arch filled and sustained by a liquid beneath, but by a mashing together of the whole crust with the formation of close folds and a thickening or swelling upward of the squeezed mass. The author walked slowly through the cut made by the Central Pacific Railroad, from the plains adjoining the bay of San Francisco through the Coast Ridge mountains to the San Joaquin plains, a distance of thirty miles. Both the sub-ranges into which the range is divided are composed wholly of crumpled strata, those of the western sub-range being crumpled in the most extraordinary manner. The sub-range nearest the bay is exceedingly complex. From measurements of the angles of dip the actual length of the folded strata is two and one-half to three times the horizontal distance through the mountain. There must have been fifteen to eighteen miles of original sea bottom crushed into six miles, with a corresponding upswelling of the whole mass.

[II.]

To anticipate inquiry and satisfy curiosity respecting the history of the author of the experiments mentioned herein, and of the book itself, his civil and military history is as follows, viz:

AUGUSTUS JAMES PLEASONTON, born in the city of Washington, in the District of Columbia, January 21st, A. D. 1808. He was the second son of Stephen Pleasonton, of the state of Delaware, and Mary Hopkins, his wife, of the county of Lancaster, state of Pennsylvania. His father, Stephen Pleasonton, entered the service of the government of the United States, in the State Department, in the year 1800, and continued to serve it till his death, which occured in the year 1854, after a service of more than fifty years. He was Fifth Auditor of the Treasury Department, Acting Commissioner of the Revenue of the United States, and Chief of the Light House Department, for many years. He was of Norman extraction.

His wife was the third daughter of John Hopkins, a substantial farmer of the county of Lancester, in the state of Pennsylvania, who for very many years represented his county in the Senate of Pennsylvania. Her ancestry was English. Their son, Augustus, was appointed a Cadet of the United States Military Academy at West Point, from the District of Columbia, July 1st, A. D. 1822, continued as such till July 1st, 1826, when he was graduated and promoted in the army, to Brevet Second Lieutenant of the Sixth Regment of Infantry July 1st, 1826, Second Lieutenant Third Artillery June 1st, 1826. Transferred to First Artillery October 24th, 1826.

Augustus James Pleasonton served in garrison at Fortress Monroe, Virginia, at the Artillery School of Practice in the years 1826 and 1827, and on Topographical duty, from June 16th, 1827, till January 17th, 1828, and from June 14th, 1828 till June 30th, 1830. Resigned his commission in the army June 30th, 1830.

HIS CIVIL HISTORY.—Counsellor at Law at Philadelphia, Penn., since the year 1832. Brigade Major in Pennsylvania Volunteer Militia in the years 1833 and 1835, Colonel of Volunteer Artillery, of Penn., from 1835 till 1845, being severely wounded July 7th, 1844, with a musket ball in the left groin, while commanding his regiment in a desperate conflict, with a formidable body of rioters, armed with muskets and cannon, in Southwark, Philadelphia county, Penn. Assistant Adjutant General and Paymaster General of the state of Pennsylvania from December 11th, 1838 to October 11th, 1839, during political disturbances at Harrisburg, Penn. President of the Harrisburg, Portsmouth, Mountjoy and Lancaster Railroad Company, of Pennsylvania, in the years 1839 and 1840.

His Military History.—Served during the Rebellion of the seceding states from the year 1861 till 1866 as Brigadier General of Pennsylvania Volunteer Militia. Appointed May 16th, 1861, under an act of the Legislature of the state of Pennsylvania, to organize and command a Volunteer Army Corps of 10,000 men of Artillery, Infantry and Cavalry, as a Home Guard for the defence of the city of Philadelphia, Penn.

Bio-
Plea es,
180 1894.
The influence of the blue
ray of the sunlight and of

Spotted something that isn't quite right?

Let us know and we'll fix it for future reprints.
And as a thank you for your diligence, we'll
send you a free Sothis book of your choice.

publisher@sothis.press

This page intentionally left blank

Made in United States
Orlando, FL
04 November 2026

90218337R00139